D1131144

FOURIER EXPANSIONS

A Collection of Formulas

Fourier Expansions

A COLLECTION OF FORMULAS

Fritz Oberhettinger

Department of Mathematics
Oregon State University
Corvallis, Oregon

ACADEMIC PRESS New York and London 1973

A Subsidiary of Harcourt Brace Jovanovich, Publishers

ACADEMIC PRESS, INC.
111 Fifth Avenue, New York, New York 10003

United Kingdom Edition published by
ACADEMIC PRESS, INC. (LONDON) LTD.
24/28 Oval Road, London NW1

Library of Congress Catalog Card Number: 79-182635

AMS (MOS) 1970 Subject Classification: 42A16

PRINTED IN THE UNITED STATES OF AMERICA

CONTENTS

PREFACE

The purpose of this monograph is to give a collection of Fourier series. Its limited scope made a number of compromises necessary. The question regarding the choice and organization of the material to be included posed certain problems. In order to preserve some consistency it seemed best to stay within the framework of what one could call the "classical" Fourier series, i.e., those of the trigonometric and their simplest generalization the Fourier–Bessel series. Thus results relating to Fourier series of generalized functions* or such series arising from Sturm–Liouville eigenvalue or integral equation problems are not included here. It was felt that such topics should be the subject of a separate treatment. An important question was which should be placed first, the Fourier series or the sum it represents. After some deliberation it was decided to opt for the first alternative. The material presented here is subdivided into five sections:

I. Series with elementary coefficients representing elementary functions

II. Series with elementary coefficients representing higher functions

III. Series with higher function coefficients representing elementary functions

IV. Series with higher function coefficients representing higher functions

V. Exponential Fourier and Fourier–Bessel series

* A few examples are given in an appendix to I.

This arrangement should be helpful in equally balancing the task of either establishing the sum function of a given Fourier series or finding the Fourier expansion of a given function. It seems apparent that a sizable amount of attention centers around results involving higher functions.

The author is not aware of the possible existence of a presentation of this subject on a similar scale and it is hoped that the contribution here will meet the requirements so often needed in applied mathematics, physics, and engineering. Since there is no lack of excellent texts concerning the subject of Fourier series no references have been given. Most of the material displayed here is widely scattered over the literature and a sizable number of results seem not to have been available before.

LIST OF NOTATION

Special Symbols

$(a)_n = a(a+1)(a+2)\cdots(a+n-1)$,
$n = 1, 2, 3, \cdots$

$(a)_0 = 1$

ϵ_n = Neumann's number, $\epsilon_0 = 1$, $\epsilon_n = 2$,
$n = 1, 2, 3, \cdots$

γ = Euler's constant

$\tau_{\nu,n}$ = nth positive root of $J_\nu(x) = 0$

k = modulus of the elliptic integrals,
$k' = (1-k^2)^{1/2}$

List of Functions*

$\operatorname{am}(z, k)$ Jacobian amplitude function of argument z and modulus k (see 2.1)

B_0 Bernoulli numbers (see 1.37)

$B_n(x)$ Bernoulli polynomials (see 1.37)

$C(x) = (2\pi)^{-1/2} \int_0^x t^{-1/2} \cos t \, dt$ Fresnel's integral

$\operatorname{Ci}(x) = -\int_x^\infty t^{-1} \cos t \, dt$ cosine integral

* The definitions are the same as in A. Erdélyi *et al.*, "Higher Transcendental Functions," 3 vols., McGraw-Hill, New York, 1953.

$\mathrm{cn}(z, k)$ Jacobian elliptic function cosine amplitude of modulus k and argument z

$C_n^\nu(z)$ Gegenbauer polynomials

$dn(z, k)$ Jacobian elliptic function delta amplitude of argument z and modulus k

$D_\nu(z)$ parabolic cylinder function

$E(k)$ complete elliptic integral of the second kind of modulus k

E_n Euler numbers (see 1.37)

$E_n(x)$ Euler polynomials

$$\mathrm{Erf}(z) = 2\pi^{-1/2} \int_0^x \exp(-t^2)\, dt = 1 - \mathrm{Erfc}(z)$$

error integrals

$$\mathrm{Erfc}(z) = 2\pi^{-1/2} \int_x^\infty \exp(-t^2)\, dt = 1 - \mathrm{Erf}(z)$$

$E_\nu(z)$ Weber's function of order ν

${}_2F_1(a, b; c; z)$ hypergeometric function

$\Gamma(z)$ Gamma function

$H_\nu^{(1),(2)}(z)$ Hankel functions of order ν

$\mathbf{H}_\nu(z)$ Struve's function of order ν

$I_\nu(z)$ modified Bessel function of order ν

$J_\nu(z)$ Bessel function of order ν

$\mathrm{J}_\nu(z)$ Anger's function of order ν

$K(k)$ complete elliptic integral of the first kind of modulus k

$K_\nu(z)$ modified Hankel function of order ν

$\psi(z)$ Euler's psi function

$P_n(x)$ Legendre's polynomials

$P_\nu^\mu(x)$ associated Legendre functions of the first kind of argument x with $-1<x<1$ and $x>1$, respectively

$Q_\nu^\mu(x)$ associated Legendre function of the second kind of argument x with $-1<x<1$ and $x>1$, respectively

$$S(x) = (2\pi)^{-1/2} \int_0^x t^{-1/2} \sin t\, dt$$ Fresnel's integral

$$\mathrm{Si}(x) = \int_0^x t^{-1/2} \sin t\, dt$$ sine integral

$\mathrm{sn}(z, k)$ Jacobian elliptic function sine amplitude of argument z and modulus k

$s_{\mu,\nu}(z)$ Lommel's function

$T_n(x) = \cos(n \arccos x)$ Chebyshev's polynomials of the first kind

$U_n(x) = (1-x^2)^{-1/2} \sin[(n+1) \arccos x]$ Chebyshev's polynomials of the second kind

$\vartheta_l(x)$ elliptic theta functions $l=0, 1, 2, 3$ (see 2.26–2.29)

$Y_\mu(z)$ Neumann function of order ν

$\mathrm{zn}(z, k)$ Jacobi zeta function of argument z and modulus k (see 2.25)

$\zeta(z, a) = \sum_1^\infty (n+a)^{-z}$ Hurwitz zeta function (see 2.51, 2.52)

INTRODUCTION

1. The Fourier Series

Let $f(x)$ be a function defined and bounded in the range $a \leq x \leq b$ satisfying the conditions

(a) $f(x)$ has only a finite number of maxima and minima in (a, b), and

(b) $f(x)$ has only a finite number of (finite) discontinuities in (a, b) and outside this range $f(x)$ is defined by the relation $f[x \pm (b-a)] = f(x)$, i.e., $f(x)$ is a periodic function of period $b-a$. If two sets of coefficients a_n and b_n (the Fourier coefficients) are defined by

$$a_n = \frac{2}{b-a} \int_a^b f(x) \cos\left(\frac{2\pi n x}{b-a}\right) dx, \qquad n = 0, 1, 2, \ldots \tag{1}$$

$$b_n = \frac{2}{b-a} \int_a^b f(x) \sin\left(\frac{2\pi n x}{b-a}\right) dx, \qquad b_n = 1, 2, 3, \ldots \tag{2}$$

then the series

$$\tfrac{1}{2}a_0 + \sum_1^\infty \left[a_n \cos\left(\frac{2n\pi x}{b-a}\right) + b_n \sin\left(\frac{2n\pi x}{b-a}\right)\right] \tag{3}$$

is called the Fourier series of $f(x)$. It converges at a point x_0 $(a \leq x_0 \leq b)$ to the sum

$$\tfrac{1}{2}[f(x_0+0)+f(x_0-0)]; \qquad f(x_0 \pm 0) = \lim_{h \to 0} f(x_0 \pm h)$$

If $b=l$ and $a=-l$, then

$$f(x) = \tfrac{1}{2}a_0 + \sum_{1}^{\infty} \{a_n \cos[n\pi(x/l)] + b_n \sin[n\pi(x/l)]\} \qquad \text{(period } 2l) \tag{4}$$

with

$$a_n = (1/l) \int_{-l}^{l} f(x) \cos[n\pi(x/l)]\, dx \tag{5}$$

$$b_n = (1/l) \int_{-l}^{l} f(x) \sin[n\pi(x/l)]\, dx \tag{6}$$

If $f(x)$ is an *even* function, then all $b_n = 0$ and

$$f(x) = \tfrac{1}{2}a_0 + \sum_{1}^{\infty} a_n \cos[n\pi(x/l)]; \qquad a_n = (2/l) \int_{0}^{l} f(x) \cos[n\pi(x/l)]\, dx \tag{7}$$

If $f(x)$ is *odd*, then all $a_n = 0$ and

$$f(x) = \sum_{1}^{\infty} b_n \sin[n\pi(x/l)]; \qquad b_n = (2/l) \int_{0}^{l} f(x) \sin[n\pi(x/l)]\, dx \tag{8}$$

The period in (7) and (8) is $2l$.

Also in exponential form

$$f(x) = \sum_{-\infty}^{\infty} c_n \exp[i2\pi nx/(b-a)]; \qquad \text{period } b-a \tag{9}$$

$$c_n = (b-a)^{-1} \int_{a}^{b} f(x) \exp[-i2\pi nx/(b-a)]\, dx \tag{10}$$

Again, if $a=-l$ and $b=l$ (period $2l$), then

$$f(x) = \sum_{-\infty}^{\infty} c_n \exp[in\pi(x/l)] \tag{11}$$

$$c_n = (2l)^{-1} \int_{-l}^{l} f(x) \exp[-in\pi(x/l)]\, dx \tag{12}$$

The case of an even or odd function $f(x)$ leads to formulas (7) and (8), respectively.

2. The Fourier Integral

The limiting case $l \to \infty$ in (11) and (12) leads to the representation of a function $f(x)$ by a Fourier integral

$$f(x) = (1/2\pi) \int_{-\infty}^{\infty} dy \int_{-\infty}^{\infty} f(t) \exp[iy(x-t)] \, dt \tag{13}$$

This is equivalent to the pair of inversion formulas (exponential Fourier transform)

$$g_e(y) = \int_{-\infty}^{\infty} f(x) e^{ixy} \, dx$$
$$f(x) = (1/2\pi) \int_{-\infty}^{\infty} g_e(y) e^{-ixy} \, dy \tag{14}$$

Again, if $f(x)$ is an even or an odd function, respectively, it follows from (14) that

$$g_c(y) = \int_0^{\infty} f(x) \cos(xy) \, dx; \qquad f(x) = (2/\pi) \int_0^{\infty} g_c(y) \cos(xy) \, dy \tag{15}$$

$$g_s(y) = \int_0^{\infty} f(x) \sin(xy) \, dx; \qquad f(x) = (2/\pi) \int_0^{\infty} g_s(y) \sin(xy) \, dy \tag{16}$$

Here $g_c(y)$ and $g_s(y)$ denote the Fourier cosine and sine transform, respectively.

3. Fourier–Bessel Series

The Fourier–Bessel series represent a generalization of the trigonometric Fourier series. They are series involving Bessel functions as terms. Let $J_\nu(x)$ be the Bessel function of the order ν and the argument x and let $\nu > -1$. Then the zeros of $J_\nu(x)$ are real and of equal absolute value. If $\tau_{\nu,n}$ denotes the nth positive root of $J_\nu(x) = 0$, then a function $f(x)$ defined in $(0, a)$ and vanishing at $x = a$ admits under certain conditions the expansion

$$f(x) = \sum_{1}^{\infty} a_n J_\nu[\tau_{\nu,n}(x/a)], \tag{17}$$

with

$$a_n = 2[aJ_{\nu+1}(\tau_{\nu,n})]^{-2} \int_0^a x f(x) J_\nu[\tau_{\nu,n}(x/a)] \, dx \tag{18}$$

For the special values $\nu=-\frac{1}{2}$ and $\nu=\frac{1}{2}$ one has

$$J_{-\frac{1}{2}}(x)=(\tfrac{1}{2}\pi x)^{-\frac{1}{2}}\cos x \qquad \text{with the zeros } \tau_{\nu,n}=(n-\tfrac{1}{2})\pi, \tag{19}$$

$$n=1,2,3,\ldots$$

$$J_{\frac{1}{2}}(x)=(\tfrac{1}{2}\pi x)^{-\frac{1}{2}}\sin x \qquad \text{with the zeros } \tau_{\nu,n}=n\pi, \tag{20}$$

The substitution $f(x)=(\frac{1}{2}\pi x)^{-\frac{1}{2}}F(x)$; $a_n=(\tau_{\nu,n}/a)^{\frac{1}{2}}A_n$ in (17) and (18) leads to

$$F(x)=\sum_1^\infty A_n\cos[(n-\tfrac{1}{2})\pi(x/a)] \quad \text{with} \quad A_n=(2/a)\int_0^a F(x)\cos[(n-\tfrac{1}{2})\pi(x/a)]\,dx \tag{21}$$

$$F(x)=\sum_1^\infty A_n\sin[n\pi(x/a)] \quad \text{with} \quad A_n=(2/a)\int_0^a F(x)\sin[n\pi(x/a)]\,dx \tag{22}$$

The periods of $F(x)$ as represented in (21) and (22) are $4a$ and $2a$, respectively. Obviously the series (21) and (22) are of the form (7) and (8).

4. The Bessel Transform

The expression of an arbitrary function $F(x)$ by means of a double integral similar to (13) but involving Bessel functions is given by Hankel's formula

$$F(x)=\int_0^\infty J_\nu(tx)t\,dt\int_0^\infty F(u)J_\nu(ut)u\,du \tag{23}$$

Equivalent to (23) is the pair of inversion formulas

$$h_\nu(y)=\int_0^\infty (xy)^{\frac{1}{2}}f(x)J_\nu(xy)\,dx; \qquad f(x)=\int_0^\infty (xy)^{\frac{1}{2}}h_\nu(y)J_\nu(xy)\,dy \tag{24}$$

which for the special cases $\nu=\pm\frac{1}{2}$ lead to formulas (15) and (16). The property $h_\nu(y)$ is called the Hankel transform of the order ν of a function $f(x)$. Also used is

$$H_\nu(y)=\int_0^\infty F(x)J_\nu(xy)\,dx; \qquad F(x)=x\int_0^\infty yH_\nu(y)J_\nu(xy)\,dy \tag{25}$$

5. Generation of Fourier Series by Means of Integral Transforms

The integrals for $g_c(y)$ and $g_s(y)$ in (15) and (16) taken over a finite interval (a, b) of integration are called the *finite cosine and sine Fourier transforms* of $f(x)$

$$g_c(y)=\int_0^a f(x)\cos(xy)\,dx; \qquad g_s(y)=\int_0^a f(x)\sin(xy)\,dy$$

If $f(x)$ is regarded as an even function of x, then by (7)

$$|f(x)| = (1/a) \sum_{0}^{\infty} \epsilon_n g_c[n(\pi/a)] \cos[n\pi(x/a)], \qquad -a<x<a \tag{26}$$

Similarly when $f(x)$ is regarded as an odd function

$$\operatorname{sgn}x\, f(x) = (2/a) \sum_{1}^{\infty} g_s[n(\pi/a)] \sin[n\pi(x/a)], \qquad -a<x<a \tag{27}$$

These formulas represent the Fourier expansion in $-a<x<a$ of a function $f(x)$ when its finite Fourier transforms are known. Similar considerations can be applied to *Poisson's summation formula*

$$\sum_{n=-\infty}^{\infty} e^{inx} G(y+nd) = (1/d) \sum_{m=-\infty}^{\infty} g_e\left(\frac{2\pi m+x}{d}\right) \exp[-i(2\pi m+x)y/d] \tag{28}$$

with

$$g_e(t) = \int_{-\infty}^{\infty} G(u) e^{iut}\, du \tag{29}$$

Here $G(u)$ is a given function and $g_e(t)$ represents its exponential Fourier transform (14); x, y, and d are real parameters.

If $y=0$, then

$$\sum_{n=-\infty}^{\infty} e^{inx} G(nd) = (1/d) \sum_{m=-\infty}^{\infty} g_e[(2\pi m+x)/d] \tag{30}$$

If $x=0$, then

$$\sum_{n=-\infty}^{\infty} G(y+nd) = (1/d) \sum_{m=-\infty}^{\infty} g_e[2\pi(m/d)] \exp[-i2\pi m(y/d)] \tag{31}$$

The left side of (31) represents a periodic function in y of period d while the right side represents its Fourier expansion in exponential form (9). Suppose now that the function $G(u)$ occurring in the left side summation of (28) is such that its exponential Fourier transform $g_e(t)$ in (29) vanishes for $a<t<b$ [i.e., $g_e(t)$ is different from zero in a finite interval only]. The sum at the right side of (28) reduces to a finite sum and

$$\sum_{n=-\infty}^{\infty} e^{inx} G(y+nd) = (1/d) \sum_{m=m_1}^{m_2} g_e[(2\pi m+x)/d] \exp[-i(y/d)(2\pi m+x)] \tag{32}$$

The limits m_1 and m_2 are then defined by

$$m_1 \geq (ad-x)/2\pi; \qquad m_2 = (bd-x)/2\pi \tag{33}$$

If b, a, d, and x are such that the equal sign holds then one half of the corresponding value in the right side summation has to be taken.

The expressions (17) and (18) for the Fourier–Bessel expansion can also be written as

$$x^{-\frac{1}{2}}f(x) = 2\sum_{1}^{\infty} \tau_{\nu,n}^{-\frac{1}{2}}[h_\nu(\tau_{\nu,n})/J_{\nu+1}^2(\tau_{\nu,n})]J_\nu(\tau_{\nu,n}x), \qquad 0<x<1 \tag{34}$$

with

$$h_\nu(y) = \int_0^1 f(t)\,(ty)^{\frac{1}{2}}J_\nu(ty)\,dt \tag{35}$$

This is the finite Hankel transform of $f(x)$.

Finally the Jacobi–Anger relations from the theory of the Bessel functions are used:

$$\cos(x\cos z) = \sum_{0}^{\infty} \epsilon_n(-1)^n J_{2n}(x)\cos(2nz) \tag{36}$$

$$\sin(x\cos z) = 2\sum_{0}^{\infty}(-1)^n J_{2n+1}(x)\cos[(2n+1)z] \tag{37}$$

Upon multiplying these relations by a suitable function of x and integrating over x from zero to infinity the results are

$$\sum_{0}^{\infty}(-1)^n\epsilon_n H_{2n}(1)\cos(2nz) = g_c(\cos z) \tag{38}$$

$$2\sum_{0}^{\infty}(-1)^n H_{2n+1}(1)\cos[(2n+1)z] = g_s(\cos z) \tag{39}$$

where g_c, g_s, and H_ν are defined in (15), (16), and (25).

I

FOURIER SERIES WITH ELEMENTARY COEFFICIENTS REPRESENTING ELEMENTARY FUNCTIONS

1.1 $$\sin^{2l}x = 2^{-2l}(2l)!\sum_{0}^{l}(-1)^n\epsilon_n\cos(2nx)/[(l+n)!(l-n)!]$$

1.2 $$\sin^{2l+1}x = 2^{-2l}(2l+1)!\sum_{0}^{l}(-1)^n\sin[(2n+1)x]/[(l+1+n)!(l-n)!]$$

1.3 $$\cos^{2l}x = 2^{-2l}(2l)!\sum_{0}^{l}[\epsilon_n\cos(2nx)]/[(l+n)!(l-n)!]$$

1.4 $$\cos^{2l+1}x = 2^{-2l}(2l+1)!\sum_{0}^{l}\cos[(2n+1)x]/[(l+1+n)!(l-n)!]$$

1.5 $$\sum_{1}^{N}\cos(nx) = \frac{\sin(\frac{1}{2}Nx)}{\sin(\frac{1}{2}x)}\cos[(N+1)\tfrac{1}{2}x] = \frac{1}{2}\left\{\frac{\sin[\frac{1}{2}(2N+1)x]}{\sin(\frac{1}{2}x)}-1\right\}$$

1.6 $$\sum_{1}^{N}\sin(nx) = [\sin(\tfrac{1}{2}Nx)/\sin(\tfrac{1}{2}x)]\sin[\tfrac{1}{2}(N+1)x]$$

$$= \tfrac{1}{2}\{\cos(\tfrac{1}{2}x)-\cos[\tfrac{1}{2}(2N+1)x]\}/\sin(\tfrac{1}{2}x)$$

1.7 $\sum_{0}^{N} \cos[(2n+1)x] = \{\sin[(N+1)x]/\sin x\} \cos[(N+1)x]$

1.8 $\sum_{0}^{N} \sin[(2n+1)x] = \{\sin[(N+1)x]/\sin x\} \sin[(N+1)x]$

1.9 $[\sin(Nx)/\sin x]^2 = \sum_{0}^{N-1} \{\sin[(2n+1)x]/\sin x\}$

$= N + 2(N-1)\cos(2x) + 2(N-2)\cos(4x) + \cdots + 2\cos[2(N-1)x]$

1.10 $\sum_{0}^{N-1} \sin(x+ny) = \sin[x+\frac{1}{2}(N-1)y][\sin(\frac{1}{2}Ny)/\sin(\frac{1}{2}y)]$

1.11 $\sum_{0}^{N-1} \cos(x+ny) = \cos[x+\frac{1}{2}(N-1)y][\sin(\frac{1}{2}Ny)/\sin(\frac{1}{2}y)]$

1.12 $\sum_{1}^{\infty} n^{-1}\sin(nx) = \frac{1}{2}\pi - \frac{1}{2}x, \quad 0<x<2\pi$

1.13 $\sum_{1}^{\infty} (-1)^n n^{-1}\sin(nx) = -\frac{1}{2}x, \quad -\pi<x<\pi$

1.14 $\sum_{1}^{\infty} n^{-1}\cos(nx) = -\log|2\sin(\frac{1}{2}x)|$

1.15 $\sum_{1}^{\infty} (-1)^n n^{-1}\cos(nx) = -\log|2\cos(\frac{1}{2}x)|$

1.16 $\sum_{0}^{\infty} (2n+1)^{-1}\sin[(2n+1)x] = \frac{1}{4}\pi \operatorname{sgn} x, \quad -\pi<x<\pi$

1.17 $\sum_{0}^{\infty} (-1)^n(2n+1)^{-1}\sin[(2n+1)x] = \frac{1}{2}\log|[\cos(\frac{1}{2}x)+\sin(\frac{1}{2}x)]/[\cos(\frac{1}{2}x)-\sin(\frac{1}{2}x)]|$

1.18 $\sum_{0}^{\infty} (2n+1)^{-1}\cos[(2n+1)x] = \frac{1}{2}\log|\cot(\frac{1}{2}x)|$

1.19 $\sum_{0}^{\infty}(-1)^n(2n+1)^{-1}\cos[(2n+1)x]=\begin{cases}\frac{1}{4}\pi, & 0<x<\frac{1}{2}\pi\\ -\frac{1}{4}\pi, & \frac{1}{2}\pi<x<\frac{3}{2}\pi\end{cases}$

1.20 $\sum_{1}^{\infty}(n+p)^{-1}\sin(nx)=(\frac{1}{2}\pi-\frac{1}{2}x)\cos(px)-\cos(px)\sum_{1}^{p}n^{-1}\sin(nx)$

$+\sin(px)\{\log[2\sin(\frac{1}{2}x)]+\sum_{1}^{p}n^{-1}\cos(nx)\},\quad p=1,2,3,\ldots;\quad 0<x<2\pi$

1.21 $\sum_{1}^{\infty}(n+p)^{-1}\cos(nx)=-\cos(px)\{\log[2\sin(\frac{1}{2}x)]+\sum_{1}^{p}n^{-1}\cos(nx)\}$

$+\sin(px)[\frac{1}{2}\pi-\frac{1}{2}x-\sum_{1}^{p}n^{-1}\sin(nx)],\quad p=1,2,3,\ldots;\quad 0<x<2\pi$

1.22 $\sum_{1}^{\infty}(n+p)^{-1}(-1)^n\sin(nx)=(-1)^{p+1}\cos(px)[\sum_{1}^{p}(-1)^n n^{-1}\sin(nx)+\frac{1}{2}x]$

$+(-1)^p\sin(px)\{\log[2\cos(\frac{1}{2}x)]+\sum_{1}^{p}(-1)^n n^{-1}\cos(nx)\},\quad p=1,2,3,\ldots;\quad -\pi<x<\pi$

1.23 $\sum_{1}^{\infty}(n+p)^{-1}(-1)^n\cos(nx)=(-1)^{p+1}\cos(px)\{\log[2\cos(\frac{1}{2}x)]+\sum_{1}^{p}(-1)^n n^{-1}\cos(nx)\}$

$-(-1)^p\sin(px)[\frac{1}{2}x+\sum_{1}^{p}(-1)^n n^{-1}\sin(nx)],\quad p=1,2,3,\ldots;\quad -\pi<x<\pi$

1.24 $\frac{4}{3}A\pi^2+B\pi+C+4A\sum_{1}^{\infty}n^{-2}\cos(nx)-(4\pi A+2B)\sum_{1}^{\infty}n^{-1}\sin(nx)$

$=Ax^2+Bx+C,\quad 0<x<2\pi,\quad A,B,C=\text{constants}$

1.25 $\sum_{1}^{\infty}n^{-2}\sin(nx)=-\int_0^x\log[2\sin(\frac{1}{2}t)]\,dt,\quad 0\leq x\leq 2\pi$

1.26 $\sum_{1}^{\infty}(-1)^n n^{-2}\sin(nx)=-\int_0^x\log[2\cos(\frac{1}{2}t)]\,dt,\quad -\pi\leq x\leq\pi$

1.27 $\sum_{1}^{\infty}n^{-2}\cos(nx)=(1/12)(3x^2-6\pi x+2\pi^2),\quad 0\leq x\leq 2\pi$

1.28 $\sum_{1}^{\infty}(-1)^n n^{-2}\cos(nx)=(1/12)(3x^2-\pi^2),\quad -\pi\leq x\leq\pi$

1.29 $\sum_{0}^{\infty} (2n+1)^{-2} \sin(nx) = -\frac{1}{2}\int_0^x \log|\tan(\frac{1}{2}t)|\,dt, \quad -\pi \leq x \leq \pi$

1.30 $\sum_{0}^{\infty} (2n+1)^{-2} \cos[(2n+1)x] = \frac{1}{8}(\pi^2 - 2\pi|x|), \quad -\pi \leq x \leq \pi$

1.31 $\sum_{0}^{\infty} (-1)^n (2n+1)^{-2} \sin[(2n+1)x] = \frac{1}{4}\pi x, \quad -\frac{1}{2}\pi \leq x \leq \frac{1}{2}\pi$

1.32 $\sum_{0}^{\infty} (-1)^n (2n+1)^{-2} \cos[(2n+1)x] = -\frac{1}{2}\int_0^{\frac{1}{2}\pi - x} \log[\tan(\frac{1}{2}t)]\,dt, \quad -\frac{1}{2}\pi \leq x \leq \frac{1}{2}\pi$

1.33 $2(-1)^{l+1}(2l)!\sum_{1}^{\infty} (2\pi n)^{-2l} \cos(2\pi n x) = B_{2l}(x), \quad 0 \leq x \leq 1, \quad l = 1, 2, 3, \ldots$

1.34 $2(-1)^{l+1}(2l+1)!\sum_{1}^{\infty} (2\pi n)^{-2l-1} \sin(2\pi n x) = B_{2l+1}(x)$

$0 \leq x \leq 1, \quad l = 1, 2, 3, \ldots; \quad 0 < x < 1, \quad l = 0$

1.35 $4(-1)^{l}(2l)!\sum_{0}^{\infty} [(2n+1)\pi]^{-2l-1} \sin[(2n+1)\pi x] = E_{2l}(x)$

$0 \leq x \leq 1, \quad l = 1, 2, 3, \ldots; \quad 0 < x < 1, \quad l = 0$

1.36 $4(-1)^{l+1}(2l+1)!\sum_{0}^{\infty} [(2n+1)\pi]^{-2l-2} \cos[(2n+1)\pi x] = E_{2l+1}(x)$

$0 \leq x \leq 1, \quad l = 0, 1, 2, \ldots$

1.37 The $B_k(x)$ and $E_k(x)$ are the Bernoulli and the Euler polynomials, respectively.

$$B_k(x) = \sum_{r=0}^{k} \binom{k}{r} B_r x^{k-r}$$

$$E_k(x) = \sum_{r=0}^{k} \binom{k}{r} 2^{-r} (x - \tfrac{1}{2})^{k-r} E_r$$

$B_0 = 1, \quad B_1 = -\frac{1}{2}, \quad B_2 = \frac{1}{6}, \quad B_4 = -\frac{1}{30}, \ldots; \quad B_{2r+1} = 0, \qquad r = 1, 2, 3, \ldots$

$E_0 = 1, \quad E_2 = -1, \quad E_4 = 5, \ldots; \quad E_{2r+1} = 0, \qquad r = 0, 1, 2, \ldots$

(Bernoulli's and Euler's numbers)

1.38 $\sum_{1}^{\infty} \{\cos[(n+1)x]/n(n+1)\} = (1-\cos x)\log[2\sin(\frac{1}{2}x)]$
$-(\frac{1}{2}\pi-\frac{1}{2}x)\sin x+\cos x, \quad 0\le x\le 2\pi$

1.39 $\sum_{1}^{\infty} \{\sin[(n+1)x]/n(n+1)\} = (\frac{1}{2}\pi-\frac{1}{2}x)(\cos x-1)$
$-\sin x\log[2\sin(\frac{1}{2}x)]+\sin x, \quad 0\le x\le 2\pi$

1.40 $\sum_{1}^{\infty} \{(-1)^n\cos[(n+1)x]/n(n+1)\} = -(1+\cos x)\log[2\cos(\frac{1}{2}x)]$
$+\frac{1}{2}x\sin x+\cos x, \quad -\pi\le x\le \pi$

1.41 $\sum_{1}^{\infty} \{(-1)^n\sin[(n+1)x]/n(n+1)\} = -\frac{1}{2}x(1+\cos x)$
$-\sin x\log[2\cos(\frac{1}{2}x)]+\sin x, \quad -\pi\le x\le \pi$

1.42 $\sum_{1}^{\infty} [(-1)^n\sin(nx)/n(n+1)] = -\sin x\log[2\sin(\frac{1}{2}x)]+(\frac{1}{2}\pi-\frac{1}{2}x)(1-\cos x), \quad 0\le x\le 2\pi$

1.43 $\sum_{1}^{\infty} [\cos(nx)/n(n+1)] = 1-(1-\cos x)\log[2\sin(\frac{1}{2}x)]-(\frac{1}{2}\pi-\frac{1}{2}x)\sin x, \quad 0\le x\le 2\pi$

1.44 $\sum_{1}^{\infty} [(-1)^n\sin(nx)/n(n+1)] = -\frac{1}{2}x(1+\cos x)+\sin x\log[2\cos(\frac{1}{2}x)], \quad -\pi\le x\le \pi$

1.45 $\sum_{1}^{\infty} [(-1)^n\cos(nx)/n(n+1)] = 1-(1+\cos x)\log[2\cos(\frac{1}{2}x)]-\frac{1}{2}x\sin x, \quad -\pi\le x\le \pi$

1.46 $(p-q)\sum_{1}^{\infty} [\cos(nx)/(n+p)(n+q)] = \cos(px)\sum_{1}^{p} n^{-1}\cos(nx)+\sin(px)\sum_{1}^{p} n^{-1}\sin(nx)$
$-\cos(qx)\sum_{1}^{q} n^{-1}\cos(nx)-\sin(qx)\sum_{1}^{q} n^{-1}\sin(nx)-2\sin[\frac{1}{2}x(p+q)]$
$\cdot\sin[\frac{1}{2}x(p-q)]\log[2\sin(\frac{1}{2}x)]-(\pi-x)\cos[\frac{1}{2}x(p+q)]\sin[\frac{1}{2}x(p-q)]$
$0\le x\le 2\pi, \quad p, q=0, 1, 2, \ldots; \quad \text{if} \quad p, q=0, \quad \sum_{1}^{0}(\ )=0$

1.47 $(p-q)\sum_1^\infty[\sin(nx)/(n+p)(n+q)]=\cos(px)\sum_1^p n^{-1}\sin(nx)-\sin(px)\sum_1^p n^{-1}\cos(nx)$

$+\sin(qx)\sum_1^q n^{-1}\cos(nx)-\cos(qx)\sum_1^q n^{-1}\sin(nx)-2\cos[\tfrac{1}{2}x(p+q)]$

$\cdot\sin[\tfrac{1}{2}x(p-q)]\log(2\sin\tfrac{1}{2}x)+(\pi-x)\sin[\tfrac{1}{2}x(p+q)]\sin[\tfrac{1}{2}x(p-q)]$

$0\le x\le 2\pi,\quad p,q=0,1,2;\quad \text{if}\quad p,q=0,\quad \sum_1^0(\ )=0$

1.48 $(p-q)\sum_1^\infty[(-1)^n\cos(nx)/(n+p)(n+q)]=(-1)^p\sum_1^p(-1)^n n^{-1}\cos[x(p-n)]$

$-(-1)^q\sum_1^q(-1)^n n^{-1}\cos[x(q-n)]+\tfrac{1}{2}x[(-1)^p\sin(px)-(-1)^q\sin(qx)]$

$+\log(2\cos\tfrac{1}{2}x)[(-1)^p\cos(px)-(-1)^q\cos(qx)]$

$-\pi\le x\le\pi,\quad p,q=0,1,2,\ldots;\quad \text{if}\quad p,q=0,\quad \sum_1^0(\ )=0$

1.49 $(p-q)\sum_1^\infty[(-1)^n\sin(nx)/(n+p)(n+q)]=-(-1)^p\sum_1^p(-1)^n n^{-1}\sin[x(p-n)]$

$+(-1)^q\sum_1^q(-1)^n n^{-1}\sin[x(q-n)]+\tfrac{1}{2}x[(-1)^p\cos(px)-(-1)^q\cos(qx)]$

$-\log(2\cos\tfrac{1}{2}x)[(-1)^p\sin(px)-(-1)^q\sin(qx)]$

$-\pi\le x\le\pi,\quad p,q=0,1,2;\quad \text{if}\quad p,q=0,\quad \sum_1^0(\ )=0$

1.50 $(r-p)\sum_1\frac{(\pm1)^n}{(n+p)(n+q)(n+r)}\begin{Bmatrix}\sin(nx)\\ \cos(nx)\end{Bmatrix},\quad p,q,r=0,1,2,\ldots$

use $\frac{r-p}{(n+p)(n+q)(n+r)}=[(n+p)(n+q)]^{-1}-[(n+q)(n+r)]^{-1}$ and 1.46–1.49

1.51 $\sum_0^\infty\frac{\epsilon_n}{b^2+n^2a^2}\cos\left(n\pi\frac{x}{l}\right)=\frac{\pi}{ab}\frac{\cosh[(\pi b/al)(l-x)]}{\sinh(\pi b/a)},\quad 0\le x\le 2l$

1.52 $\sum_0^\infty\frac{(-1)^n\epsilon_n}{b^2+n^2a^2}\cos\left(n\pi\frac{x}{l}\right)=\frac{\pi}{ab}\frac{\cosh[(\pi b/al)x]}{\sinh(\pi b/a)},\quad -l\le x\le l$

1.53 $$\sum_{1}^{\infty} \frac{n}{b^2+n^2a^2} \sin\left(n\pi \frac{x}{l}\right) = \frac{\pi}{2a^2} \frac{\sinh[(\pi b/al)(l-x)]}{\sinh(\pi b/a)}, \quad 0<x<2l$$

1.54 $$\sum_{1}^{\infty} \frac{(-1)^n n}{b^2+n^2a^2} \sin\left(n\pi \frac{x}{l}\right) = -\frac{\pi}{2a^2} \frac{\sinh[(\pi b/al)x]}{\sinh(\pi b/a)}, \quad -l<x<l$$

1.55 $$\sum_{0}^{\infty} [b^2+(n+\tfrac{1}{2})^2a^2]^{-1} \cos\left[\left(n+\frac{1}{2}\right)\pi \frac{x}{l}\right] = \frac{\pi}{2ab} \frac{\sinh[(\pi b/al)(l-x)]}{\cosh(\pi b/a)}, \quad 0\leq x\leq 2l$$

1.56 $$\sum_{0}^{\infty} \frac{(n+\frac{1}{2})}{b^2+(n+\frac{1}{2})^2a^2} \sin\left[\left(n+\frac{1}{2}\right)\pi \frac{x}{l}\right] = \frac{\pi}{2a^2} \frac{\cosh[(\pi b/al)(l-x)]}{\cos(\pi b/a)}, \quad 0<x<2l$$

1.57 $$\sum_{0}^{\infty} \frac{(-1)^n}{b^2+(n+\frac{1}{2})^2a^2} \sin\left[\left(n+\frac{1}{2}\right)\pi \frac{x}{l}\right] = \frac{\pi}{2ab} \frac{\sinh[(\pi b/al)x]}{\cosh(\pi b/a)}, \quad -l\leq x\leq l$$

1.58 $$\sum_{0}^{\infty} \frac{(-1)^n(n+\frac{1}{2})}{b^2+(n+\frac{1}{2})^2a^2} \cos\left[\left(n+\frac{1}{2}\right)\pi \frac{x}{l}\right] = \frac{\pi}{2a^2} \frac{\cosh[(\pi b/al)x]}{\cosh(\pi b/a)}, \quad -l<x<l$$

1.59 $$\sum_{0}^{\infty} \frac{\epsilon_n}{n^2a^2-b^2} \cos\left(n\pi \frac{x}{l}\right) = -\frac{\pi}{ab} \frac{\cos[(\pi b/al)(l-x)]}{\sin(\pi b/a)}, \quad 0\leq x\leq 2l$$

1.60 $$\sum_{0}^{\infty} \frac{(-1)^n\epsilon_n}{n^2a^2-b^2} \cos\left(n\pi \frac{x}{l}\right) = -\frac{\pi}{ab} \frac{\cos[(\pi b/al)x]}{\sin(\pi b/a)}, \quad -l\leq x\leq l$$

1.61 $$\sum_{1}^{\infty} \frac{n}{n^2a^2-b^2} \sin\left(n\pi \frac{x}{l}\right) = \frac{\pi}{2a^2} \frac{\sin[(\pi b/al)(l-x)]}{\sin(\pi b/a)}, \quad 0<x<2l$$

1.62 $$\sum_{1}^{\infty} \frac{(-1)^n n}{n^2a^2-b^2} \sin\left(n\pi \frac{x}{l}\right) = -\frac{\pi}{2a^2} \frac{\sin[(\pi b/al)x]}{\sin(\pi b/a)}, \quad -l<x<l$$

1.63 $$\sum_{0}^{\infty} [(n+\tfrac{1}{2})^2a^2-b^2]^{-1} \cos\left[\left(n+\frac{1}{2}\right)\pi \frac{x}{l}\right] = \frac{\pi}{2ab} \frac{\sin[(\pi b/al)(l-x)]}{\cos(\pi b/a)}, \quad 0\leq x\leq 2l$$

1.64 $$\sum_{0}^{\infty} \frac{(n+\frac{1}{2})}{(n+\frac{1}{2})^2a^2-b^2} \sin\left[\left(n+\frac{1}{2}\right)\pi \frac{x}{l}\right] = \frac{\pi}{2a^2} \frac{\cos[(\pi b/al)(l-x)]}{\cos(\pi b/a)}, \quad 0<x<2l$$

1.65 $$\sum_{0}^{\infty} \frac{(-1)^n}{(n+\frac{1}{2})^2a^2-b^2} \sin\left[\left(n+\frac{1}{2}\right)\pi \frac{x}{l}\right] = \frac{\pi}{2ab} \frac{\sin[(\pi b/al)x]}{\cos(\pi b/a)}, \quad -l\leq x\leq l$$

1.66 $$\sum_0^\infty \frac{(-1)^n(n+\frac{1}{2})}{(n+\frac{1}{2})^2a^2-b^2}\cos\left[\left(n+\frac{1}{2}\right)\pi\frac{x}{l}\right]=\frac{\pi}{2a^2}\frac{\cos[(\pi b/al)x]}{\cos(\pi b/a)},\quad -l<x<l$$

1.67 $$\sum_0^\infty \frac{\epsilon_n(-1)^n}{b^2+n^2a^2}\left[\frac{b}{a}\cos\left(n\pi\frac{x}{l}\right)-n\sin\left(n\pi\frac{x}{l}\right)\right]=\frac{\pi}{a^2}\frac{\exp[(\pi b/al)x]}{\sinh[\pi(b/a)]},\quad -l<x<l$$

1.68 $$\sum_0^\infty \frac{\epsilon_n(-1)^n}{b^2+n^2a^2}\left[\frac{b}{a}\cos\left(n\pi\frac{x}{l}\right)+n\sin\left(n\pi\frac{x}{l}\right)\right]=\frac{\pi}{a^2}\frac{\exp[-(\pi b/al)x]}{\sinh[\pi(b/a)]},\quad -l<x<l$$

1.69 $$\sum_0^\infty \frac{(-1)^n}{b^2+(n+\frac{1}{2})^2a^2}\left\{\frac{b}{a}\sin\left[\left(n+\frac{1}{2}\right)\pi\frac{x}{l}\right]+\left(n+\frac{1}{2}\right)\cos\left[\left(n+\frac{1}{2}\right)\pi\frac{x}{l}\right]\right\}$$

$$=\frac{\pi}{2a^2}\frac{\exp[(\pi b/al)x]}{\cosh(\pi b/a)},\quad -l<x<l\quad \text{(period } 4l\text{)}$$

1.70 $$\sum_0^\infty \frac{(-1)^n}{b^2+(n+\frac{1}{2})^2a^2}\left\{\frac{b}{a}\sin\left[\left(n+\frac{1}{2}\right)\pi\frac{x}{l}\right]-\left(n+\frac{1}{2}\right)\cos\left[\left(n+\frac{1}{2}\right)\pi\frac{x}{l}\right]\right\}$$

$$=-\frac{\pi}{2a^2}\frac{\exp[-(\pi b/al)x]}{\cosh(\pi b/a)},\quad -l<x<l\quad \text{(period } 4l\text{)}$$

1.71 $$\sum_0^\infty \frac{\epsilon_n}{b^2+n^2a^2}\left[\frac{b}{a}\cos\left(n\pi\frac{x}{l}\right)+n\sin\left(n\pi\frac{x}{l}\right)\right]=\frac{\pi}{a^2}\frac{\exp[(\pi b/al)(l-x)]}{\sinh(\pi b/a)},\quad 0<x<2l$$

1.72 $$\sum_0^\infty \frac{\epsilon_n}{b^2+n^2a^2}\left[\frac{b}{a}\cos\left(n\pi\frac{x}{l}\right)-n\sin\left(n\pi\frac{x}{l}\right)\right]=\frac{\pi}{a^2}\frac{\exp[-(\pi b/al)(l-x)]}{\sinh(\pi b/a)},\quad 0<x<2l$$

1.73 $$\sum_0^\infty [b^2+(n+\tfrac{1}{2})^2a^2]^{-1}\left\{\frac{b}{a}\cos\left[\left(n+\frac{1}{2}\right)\pi\frac{x}{l}\right]+\left(n+\frac{1}{2}\right)\sin\left[\left(n+\frac{1}{2}\right)\pi\frac{x}{l}\right]\right\}$$

$$=\frac{\pi}{2a^2}\frac{\exp[(\pi b/al)(l-x)]}{\cosh(\pi b/a)},\quad 0<x<2l$$

1.74 $$\sum_0^\infty [b^2+(n+\tfrac{1}{2})^2a^2]^{-1}\left\{\frac{b}{a}\cos\left[\left(n+\frac{1}{2}\right)\pi\frac{x}{l}\right]-\left(n+\frac{1}{2}\right)\sin\left[\left(n+\frac{1}{2}\right)\pi\frac{x}{l}\right]\right\}$$

$$=-\frac{\pi}{2a^2}\frac{\exp[-(\pi b/al)(l-x)]}{\cosh(\pi b/a)},\quad 0<x<2l$$

1.75 $\sum_{1}^{\infty} z^n \sin(nx) = z \sin x/(1-2z \cos x+z^2), \quad |z|<1$

1.76 $\sum_{0}^{\infty} z^n \cos(nx) = (1-z \cos x)/(1-2z \cos x+z^2), \quad |z|<1$

1.77 $\sum_{1}^{\infty} n^{-1}z^n \sin(nx) = \arctan[z \sin x/(1-z \cos x)], \quad |z|\leq 1$

1.78 $\sum_{1}^{\infty} n^{-1} z^n \cos(nx) = -\frac{1}{2} \log(1-2z \cos x+z^2), \quad |z|\leq 1$

1.79 $\sum_{0}^{\infty} z^{2n+1} \sin[(2n+1)x] = z \sin x(1+z^2)/[(1+z^2)^2-4z^2 \cos^2 x], \quad |z|<1$

1.80 $\sum_{0}^{\infty} z^{2n+1} \cos[(2n+1)x] = z \cos x(1-z^2)/[(1+z^2)^2-4z^2 \cos^2 x], \quad |z|<1$

1.81 $\sum_{0}^{\infty} (-1)^n z^{2n+1} \sin[(2n+1)x] = z \sin x(1-z^2)/[(1+z^2)^2-4z^2 \sin^2 x], \quad |z|<1$

1.82 $\sum_{0}^{\infty} (-1)^n z^{2n+1} \cos[(2n+1)x] = z \cos x(1+z^2)/[(1+z^2)^2-4z^2 \sin^2 x], \quad |z|<1$

1.83 $\sum_{0}^{\infty} (2n+1)^{-1} z^{2n+1} \sin[(2n+1)x] = \frac{1}{2} \arctan[2z \sin x/(1-z^2)], \quad |z|\leq 1$

1.84 $\sum_{0}^{\infty} (2n+1)^{-1} z^{2n+1} \cos[(2n+1)x] = \frac{1}{4} \log[(1+2z \cos x+z^2)/(1-2z \cos x+z^2)], \quad |z|\leq 1$

1.85 $\sum_{0}^{\infty} (-1)^n (2n+1)^{-1} z^{2n+1} \sin[(2n+1)x] = \frac{1}{4} \log[(1+2z \sin x+z^2)/(1-2z \sin x+z^2)],$

$|z|\leq 1$

1.86 $\sum_{0}^{\infty} (-1)^n (2n+1)^{-1} z^{2n+1} \cos[(2n+1)x] = \frac{1}{2} \arctan[2z \cos x/(1-z^2)], \quad |z|\leq 1$

1.87 $\sum_{1}^{\infty} e^{-nt}\sin(nx)=\frac{1}{2}[\sin x/(\cosh t-\cos x)]$

1.88 $\sum_{1}^{\infty} (-1)^n e^{-nt}\sin(nx)=-\frac{1}{2}[\sin x/(\cosh t+\cos x)]$

1.89 $\sum_{0}^{\infty} \epsilon_n e^{-nt}\cos(nx)=\sinh t/(\cosh t-\cos x)$

1.90 $\sum_{0}^{\infty} (-1)^n\epsilon_n e^{-nt}\cos(nx)=\sinh t/(\cosh t+\cos x)$

1.91 $\sum_{0}^{\infty} \exp[-(2n+1)t]\sin[(2n+1)x]=\sin x\cosh t/[\cosh(2t)-\cos(2x)]$

1.92 $\sum_{0}^{\infty} (-1)^n\exp[-(2n+1)t]\sin[(2n+1)x]=\sin x\sinh t/[\cosh(2t)+\cos(2x)]$

1.93 $\sum_{0}^{\infty} \exp[-(2n+1)t]\cos[(2n+1)x]=\cos x\sinh t/[\cosh(2t)-\cos(2x)]$

1.94 $\sum_{0}^{\infty} (-1)^n\exp[-(2n+1)t]\cos[(2n+1)x]=\cos x\cosh t/[\cosh(2t)+\cos(2x)]$

1.95 $\sum_{1}^{\infty} n^{-1}e^{-nt}\sin(nx)=\arctan[\sin x/(e^t-\cos x)]$

1.96 $\sum_{1}^{\infty} n^{-1}e^{-nt}\cos(nx)=\frac{1}{2}t-\frac{1}{2}\log[2(\cosh t-\cos x)]$

1.97 $\sum_{1}^{\infty} (-1)^n n^{-1}e^{-nt}\sin(nx)=-\arctan[\sin x/(e^t+\cos x)]$

1.98 $\sum_{1}^{\infty} (-1)^n n^{-1}e^{-nt}\cos(nx)=\frac{1}{2}t-\frac{1}{2}\log[2(\cosh t+\cos x)]$

1.99 $\sum_{0}^{\infty} (2n+1)^{-1}\exp[-(2n+1)t]\sin[(2n+1)x]=\frac{1}{2}\arctan(\sin x/\sinh t)$

1.100 $\sum_{0}^{\infty} (2n+1)^{-1} \exp[-(2n+1)t] \cos[(2n+1)x] = \frac{1}{4} \log[(\cosh t + \cos x)/(\cosh t - \cos x)]$

1.101 $\sum_{0}^{\infty} (-1)^n (2n+1)^{-1} \exp[-(2n+1)t] \sin[(2n+1)x]$

$= \frac{1}{4} \log[(\cosh t + \sin x)/(\cosh t - \sin x)]$

1.102 $\sum_{0}^{\infty} (-1)^n (2n+1)^{-1} \exp[-(2n+1)t] \cos[(2n+1)x] = \frac{1}{2} \arctan(\cos x/\sinh t)$

1.103 $\sum_{1}^{\infty} (z^n/n!) \sin(nx) = \exp(z \cos x) \sin(z \sin x)$

1.104 $\sum_{0}^{\infty} (z^n/n!) \cos(nx) = \exp(z \cos x) \cos(z \sin x)$

1.105 $\sum_{0}^{\infty} [z^{2n+1}/(2n+1)!] \sin[(2n+1)x] = \sin(z \sin x) \cosh(z \cos x)$

1.106 $\sum_{1}^{\infty} [z^{2n}/(2n)!] \sin(2nx) = \sin(z \sin x) \sinh(z \cos x)$

1.107 $\sum_{0}^{\infty} [z^{2n+1}/(2n+1)!] \cos[(2n+1)x] = \cos(z \sin x) \sinh(z \cos x)$

1.108 $\sum_{0}^{\infty} [z^{2n}/(2n)!] \cos(2nx) = \cos(z \sin x) \cosh(z \cos x)$

1.109 $\sum_{0}^{\infty} (-1)^n [z^{2n+1}/(2n+1)!] \sin[(2n+1)x] = \sinh(z \sin x) \cosh(z \cos x)$

1.110 $\sum_{1}^{\infty} (-1)^n [z^{2n}/(2n)!] \sin(2nx) = -\sinh(z \sin x) \sin(z \cos x)$

1.111 $\sum_{0}^{\infty} (-1)^n [z^{2n+1}/(2n+1)!] \cos[(2n+1)x] = \cosh(z \sin x) \sin(z \cos x)$

1.112 $\sum_{0}^{\infty} (-1)^n[z^{2n}/(2n)!]\cos(2nx) = \cosh(z \sin x) \cos(z \cos x)$

1.113 $\sum_{1}^{\infty} n^{-1} \sin(nx) \sin(ny) e^{-nt} = \frac{1}{4} \log\{[\cosh t - \cos(x+y)]/[\cosh t - \cos(x-y)]\}$

1.114 $\sum_{1}^{\infty} n^{-1} \sin(nx) \cos(ny) e^{-nt}$

$= \frac{1}{2} \arctan[(2e^t \sin x \cos y - 2 \sin x \cos x)/(e^{2t} - 2e^t(\cos x \cos y + \cos 2x)]$

1.115 $\sum_{1}^{\infty} n^{-1} \cos(nx) \cos(ny) e^{-nt} = \frac{1}{2}t - \frac{1}{4} \log\{4[\cosh t - \cos(x+y)][\cosh t - \cos(x-y)]\}$

Appendix: Some Results Involving Generalized Functions

The unit step function $U(t)$ and the "delta" function $\delta(t)$ are defined by

$$U(t) = 0, \qquad t < 0$$

$$U(t) = 1, \qquad t > 0$$

$$\delta(t) = U'(t)$$

Some formerly listed results and some of their generalizations can be expressed in terms of the unit step function $U(t)$. Also, some formal nonconvergent Fourier series can be interpreted as an infinite number of rows of delta functions. With the definition of two sets of four functions, $g_m(x)$, $h_m(x)$, $m = 1, 2, 3, 4$:

$$g_1(x) = \sum_{0}^{\infty} U(x - 2na + y) - \sum_{0}^{\infty} U(x - 2na - y)$$

$$g_2(x) = \sum_{0}^{\infty} U(x - 2na + y) + \sum_{0}^{\infty} U(x - 2na - y)$$

$$g_3(x) = \sum_{0}^{\infty} (-1)^n U(x - 2na + y) - \sum_{0}^{\infty} (-1)^n U(x - 2na - y)$$

$$g_4(x) = \sum_{0}^{\infty} (-1)^n U(x - 2na + y) + \sum_{0}^{\infty} (-1)^n U(x - 2na - y)$$

$$h_1(x) = \sum_0^\infty \delta(x-2na+y) - \sum_0^\infty \delta(x-2na-y)$$

$$h_2(x) = \sum_0^\infty \delta(x-2na+y) + \sum_0^\infty \delta(x-2na-y)$$

$$h_3(x) = \sum_0^\infty (-1)^n\delta(x-2na+y) - \sum_0^\infty (-1)^n\delta(x-2na-y)$$

$$h_4(x) = \sum_0^\infty (-1)^n\delta(x-2na+y) + \sum_0^\infty (-1)^n\delta(x-2na-y)$$

[The $g_m(x)$ and $h_m(x)$ are regarded as functions of $x>0$ (this is no loss of generality since the terms in the Fourier series involved are either even or odd functions of x) with parameter y $(-a<y<a)$.] One has the following formulas.

1.116 $\sum_1^\infty n^{-1} \sin[n\pi(x/a)] = -\tfrac{1}{2}\pi[1+(x/a)-2\sum_0^\infty U(x-2na)]$

1.117 $\sum_1^\infty (-1)^n n^{-1} \sin[n\pi(x/a)] = -\tfrac{1}{2}\pi\{(x/a)-2\sum_0^\infty U[x-(2n+1)a]\}$

1.118 $\sum_0^\infty (2n+1)^{-1} \sin[(2n+1)(\pi x/2a)] = -\tfrac{1}{2}\pi[\tfrac{1}{2}-\sum_0^\infty (-1)^n U(x-2na)]$

1.119 $\sum_0^\infty (-1)^n(2n+1)^{-1} \cos[(2n+1)(\pi x/2a)] = \tfrac{1}{2}\pi\{\tfrac{1}{2}-\sum_0^\infty (-1)^n U[x-(2n+1)a]\}$

1.120 $\sum_1^\infty n^{-1} \sin[n\pi(y/a)] \cos[n\pi(x/a)] = \tfrac{1}{2}\pi[g_1(x)-(y/a)]$

1.121 $\sum_1^\infty n^{-1} \cos[n\pi(y/a)] \sin[n\pi(x/a)] = \tfrac{1}{2}\pi[g_2(x)-1-(x/a)]$

1.122 $\sum_0^\infty (2n+1)^{-1} \sin[(2n+1)(\pi y/2a)] \cos[(2n+1)(\pi x/2a)] = \tfrac{1}{4}\pi g_3(x)$

1.123 $\sum_0^\infty (2n+1)^{-1} \cos[(2n+1)(\pi y/2a)] \sin[(2n+1)(\pi x/2a)] = \tfrac{1}{4}\pi[g_4(x)-1]$

1.124 $\sum_{1}^{\infty} (-1)^n n^{-1} \sin[n\pi(y/a)] \cos[n\pi(x/a)] = \frac{1}{2}\pi[g_1(x-a) - (y/a)]$

1.125 $\sum_{1}^{\infty} (-1)^n n^{-1} \cos[n\pi(y/a)] \sin[n\pi(x/a)] = \frac{1}{2}\pi[g_2(x-a) - (x/a)]$

1.126 $\sum_{0}^{\infty} (-1)^n (2n+1)^{-1} \sin[(2n+1)(\pi y/2a)] \sin[(2n+1)(\pi x/2a)] = \frac{1}{4}\pi g_3(x-a)$

1.127 $\sum_{0}^{\infty} (-1)^n (2n+1)^{-1} \cos[(2n+1)(\pi y/2a)] \cos[(2n+1)(\pi x/2a)] = \frac{1}{4}\pi[1 - g_4(x-a)]$

The formulas 1.116–1.119 are equivalent to 1.12, 1.13, 1.16, and 1.19. It should be pointed out that the periodicity in x (equal to $2a$) is preserved for both sides of the formulas before. Nonconvergent Fourier series representing an infinite number of rows of delta functions can be obtained by formally differentiating term by term the results 1.116–1.127.

1.128 $\sum_{0}^{\infty} \epsilon_n \cos[n\pi(x/a)] = 2a \sum_{0}^{\infty} \delta(x-2na)$

1.129 $\sum_{0}^{\infty} (-1)^n \epsilon_n \cos[n\pi(x/a)] = 2a \sum_{0}^{\infty} \delta[x-(2n+1)a]$

1.130 $\sum_{0}^{\infty} \cos[(2n+1)(\pi x/2a)] = a \sum_{0}^{\infty} (-1)^n \delta(x-2na)$

1.131 $\sum_{0}^{\infty} (-1)^n \sin[(2n+1)(\pi x/2a)] = a \sum_{0}^{\infty} (-1)^n \delta[x-(2n+1)a]$

1.132 $\sum_{1}^{\infty} \sin[n\pi(y/a)] \sin[n\pi(x/a)] = -\frac{1}{2}a h_1(x)$

1.133 $\sum_{1}^{\infty} \cos[n\pi(y/a)] \cos[n\pi(x/a)] = \frac{1}{2}a[h_2(x) - a^{-1}]$

1.134 $\sum_{0}^{\infty} \sin[(2n+1)(\pi y/2a)] \sin[(2n+1)(\pi x/2a)] = -\frac{1}{2}a h_3(x)$

1.135 $\sum_{0}^{\infty} \cos[(2n+1)(\pi y/2a)] \cos[(2n+1)(\pi x/2a)] = \frac{1}{2}ah_4(x)$

1.136 $\sum_{1}^{\infty} (-1)^n \sin[n\pi(y/a)] \sin[n\pi(x/a)] = -\frac{1}{2}ah_1(x-a)$

1.137 $\sum_{1}^{\infty} (-1)^n \cos[n\pi(y/a)] \cos[n\pi(x/a)] = \frac{1}{2}a[h_2(x-a) - a^{-1}]$

1.138 $\sum_{0}^{\infty} (-1)^n \sin[(2n+1)(\pi y/2a)] \cos[(2n+1)(\pi x/2a)] = \frac{1}{2}ah_3(x-a)$

1.139 $\sum_{0}^{\infty} (-1)^n \cos[(2n+1)(\pi y/2a)] \cos[(2n+1)(\pi x/2a)] = -\frac{1}{2}ah_4(x-a)$

II

FOURIER SERIES WITH ELEMENTARY COEFFICIENTS REPRESENTING HIGHER FUNCTIONS

Formulas 2.1–2.25 involve the Jacobian elliptic functions $\mathrm{am}(z, k)$, $\mathrm{sn}(z, k)$, $\mathrm{cn}(z, k)$, $\mathrm{dn}(z, k)$, and $\mathrm{zn}(z, k)$. The parameter k is omitted in these formulas, i.e., $\mathrm{sn}z=\mathrm{sn}(z, k)$, etc.

In 2.26–2.37 $\vartheta_l(z) = \vartheta_l(z, q)$ are the Jacobian theta functions defined in 2.26–2.29. Finally $K(k)$ and $E(k)$ are the complete elliptic integrals of the first and second kind of the modulus k, respectively; $k'^2=1-k^2$. Also

$$q=\exp(i\pi\tau), \qquad \tau=i(K'/K); \qquad \mathrm{Im}\tau>0; \qquad q=\exp[-\pi(K'/K)]$$

Furthermore

$$Q_0=\prod_1^\infty (1-q^{2n}), \qquad \log Q_0=-\sum_1^\infty [n^{-1}q^{2n}/(1-q^{2n})],$$

$$q^n/(1-q^{2n})=\tfrac{1}{2}\{\sinh[n\pi(K'/K)]\}^{-1};$$

$$q^n/(1+q^{2n})=\tfrac{1}{2}\{\cosh[n\pi(K'/K)]\}^{-1}, \qquad q^{n+\frac{1}{2}}/(1-q^{2n+1})=\tfrac{1}{2}\{\sinh[(n+\tfrac{1}{2})\pi(K'/K)]\}^{-1};$$

$$[q^{n+\frac{1}{2}}/(1+q^{2n+1})]=\tfrac{1}{2}\{\cosh[(n+\tfrac{1}{2})(\pi K'/K)]\}^{-1}$$

Restrictions In formulas 2.1–2.4, 2.7, 2.9, 2.11, 2.14, 2.15, 2.17, 2.18, 2.20, 2.21, 2.23, 2.25, 2.30, 2.31

$$|\,\mathrm{Im}z\,|<\tfrac{1}{2}\pi\ \mathrm{Im}\tau$$

In 2.5, 2.6, 2.8, 2.10, 2.12, 2.13, 2.16, 2.22, 2.24, 2.32, 2.33

$$|\operatorname{Im} z| < \pi \operatorname{Im} \tau$$

In 2.19

$$|\operatorname{Im} z| < 2\pi \operatorname{Im} \tau$$

2.1 $\sum_1^\infty [n^{-1}q^n/(1+q^{2n})] \sin(2nz) = \frac{1}{2} \operatorname{am}[2K(z/\pi)] - \frac{1}{2}z$

2.2 $\sum_1^\infty [q^{n-\frac{1}{2}}/(1-q^{2n-1})] \sin[(2n-1)z] = (kK/2\pi) \operatorname{sn}(2Kz/\pi)$

2.3 $\sum_1^\infty [q^{n-\frac{1}{2}}/(1+q^{2n-1})] \cos[(2n-1)z] = (kK/2\pi) \operatorname{cn}(2Kz/\pi)$

2.4 $\sum_1^\infty [q^n/(1+q^{2n})] \cos(2nz) = (K/2\pi) \operatorname{dn}(2Kz/\pi) - \frac{1}{4}$

2.5 $\sum_1^\infty [q^{2n-1}/(1-q^{2n-1})] \sin[(2n-1)z] = \dfrac{K}{2\pi \operatorname{sn}(2Kz/\pi)} - (4 \sin z)^{-1}$

2.6 $\sum_1^\infty [(-1)^n q^{2n-1}/(1+q^{2n-1})] \cos[(2n-1)z] = \dfrac{k'K}{2\pi \operatorname{cn}(2Kz/\pi)} - (4 \cos z)^{-1}$

2.7 $\sum_1^\infty [(-1)^n q^n/(1+q^{2n})] \cos(2nz) = \dfrac{k'K}{2\pi \operatorname{dn}(2Kz/\pi)} - \dfrac{1}{4}$

2.8 $\sum_1^\infty [(-1)^n q^{2n}/(1+q^{2n})] \sin(2nz) = \dfrac{k'K \operatorname{sn}(2Kz/\pi)}{2\pi \operatorname{cn}(2Kz/\pi)} - \dfrac{1}{4} \tan z$

2.9 $\sum_1^\infty [(-1)^n q^{n-\frac{1}{2}}/(1+q^{2n-1})] \sin[(2n-1)z] = -\dfrac{kk'K \operatorname{sn}(2Kz/\pi)}{2\pi \operatorname{dn}(2Kz/\pi)}$

2.10 $\sum_1^\infty [q^{2n}/(1+q^{2n})] \sin(2nz) = -\dfrac{K \operatorname{cn}(2Kz/\pi)}{2\pi \operatorname{sn}(2Kz/\pi)} + \dfrac{1}{4} \cot z$

2.11 $\sum_1^\infty [(-1)^n q^{n-\frac{1}{2}}/(1-q^{2n-1})] \cos[(2n-1)z] = -\dfrac{kK \operatorname{sn}(2Kz/\pi)}{2\pi \operatorname{dn}(2Kz/\pi)}$

2.12 $\sum_{1}^{\infty} [q^{2n-1}/(1+q^{2n-1})] \sin[(2n-1)z] = -\frac{K \operatorname{dn}(2Kz/\pi)}{2\pi \operatorname{sn}(2Kz/\pi)} + (4 \sin z)^{-1}$

2.13 $\sum_{1}^{\infty} [(-1)^n q^{2n-1}/(1-q^{2n-1})] \cos[(2n-1)z] = -\frac{K \operatorname{dn}(2Kz/\pi)}{2\pi \operatorname{cn}(2Kz/\pi)} + (4 \cos z)^{-1}$

2.14 $\sum_{1}^{\infty} [q^n/(1+q^n)] \sin(2nz) = -\frac{K \operatorname{cn}(2Kz/\pi) \operatorname{dn}(2Kz/\pi)}{2\pi \operatorname{sn}(2Kz/\pi)} + \frac{1}{4} \cot z$

2.15 $\sum_{1}^{\infty} \{q^n/[1+(-1)^n q^n]\} \sin(2nz) = \frac{K \operatorname{sn}(2Kz/\pi) \operatorname{dn}(2Kz/\pi)}{2\pi \operatorname{cn}(2Kz/\pi)} - \frac{1}{4} \tan z$

2.16 $\sum_{1}^{\infty} [q^{2n-1}/(1-q^{4n-2})] \sin[(2n-1)z] = \frac{k^2 K \operatorname{sn}(Kz/\pi) \operatorname{cn}(Kz/\pi)}{4\pi \operatorname{dn}(Kz/\pi)}$

2.17 $\sum_{1}^{\infty} [(-1)^n q^n/(1-q^n)] \sin(2nz) = \frac{k'^2 K \operatorname{sn}(2Kz/\pi)}{2\pi \operatorname{cn}(2Kz/\pi) \operatorname{dn}(2Kz/\pi)} - \frac{1}{4} \tan z$

2.18 $\sum_{1}^{\infty} \{(-1)^n q^n/[1+(-1)^n q^n]\} \sin(2nz) = -\frac{K \operatorname{cn}(2Kz/\pi)}{2\pi \operatorname{sn}(2Kz/\pi) \operatorname{dn}(2Kz/\pi)} + \frac{1}{4} \cot z$

2.19 $\sum_{1}^{\infty} [q^{4n-2}/(1-q^{4n-2})] \sin[(2n-1)z] = \frac{K \operatorname{dn}(Kz/\pi)}{4\pi \operatorname{sn}(Kz/\pi) \operatorname{cn}(Kz/\pi)} - \frac{1}{4 \sin z}$

2.20 $\sum_{1}^{\infty} [n^{-1} q^n/(1+q^n)] \cos(2nz) = \frac{1}{2} \log \operatorname{sn}(2Kz/\pi) - \frac{1}{2} \log(2q^{\frac{1}{4}} k^{-\frac{1}{2}} \sin z)$

2.21 $\sum_{1}^{\infty} \{n^{-1} q^n/[1+(-q)^n]\} \cos(2nz) = \frac{1}{2} \log \operatorname{cn}(2Kz/\pi) - \frac{1}{2} \log[2q^{\frac{1}{4}}(k'/k)^{\frac{1}{2}} \cos z]$

2.22 $\sum_{1}^{\infty} [(2n-1)^{-1} q^{2n-1}/(1-q^{4n-2})] \cos[(4n-2)z] = -\frac{1}{8} \log k' + \frac{1}{4} \log \operatorname{dn}(2Kz/\pi)$

2.23 $\sum_{1}^{\infty} [nq^n/(1-q^{2n})] \cos(2nz) = (K^2/2\pi^2) \operatorname{dn}^2(2Kz/\pi) - (KE/2\pi^2)$

2.24 $\sum_{1}^{\infty} [nq^{2n}/(1-q^{2n})] \cos(2nz) = -\frac{K^2}{2\pi^2 \operatorname{sn}^2(2Kz/\pi)} + (8 \sin^2 z)^{-1} + \frac{K(K-E)}{2\pi^2}$

2.25 $\sum_{1}^{\infty} [q^n/(1-q^{2n})] \sin(2nz) = (K/2\pi) \operatorname{zn}[(2K/\pi)z]$

2.26 $\sum_{0}^{\infty} (-1)^n \epsilon_n q^{n^2} \cos(2nz) = \vartheta_0(z/\pi)$

2.27 $\sum_{0}^{\infty} \epsilon_n q^{n^2} \cos(2nz) = \vartheta_3(z/\pi)$

2.28 $\sum_{0}^{\infty} (-1)^n q^{(n+\frac{1}{2})^2} \sin[(2n+1)z] = \frac{1}{2}\vartheta_1(z/\pi)$

2.29 $\sum_{0}^{\infty} q^{(n+\frac{1}{2})^2} \cos[(2n+1)z] = \frac{1}{2}\vartheta_2(z/\pi)$

2.30 $\sum_{1}^{\infty} [n^{-1}q^n/(1-q^{2n})] \cos(2nz) = \frac{1}{2} \log Q_0 - \frac{1}{2} \log \vartheta_0(z/\pi)$

2.31 $\sum_{1}^{\infty} [n^{-1}q^n(-1)^n/(1-q^{2n})] \cos(2nz) = \frac{1}{2} \log Q_0 - \frac{1}{2} \log \vartheta_3(z/\pi)$

2.32 $\sum_{1}^{\infty} [n^{-1}q^{2n}/(1-q^{2n})] \cos(2nz) = \frac{1}{2} \log(2Q_0) + \frac{1}{8} \log q + \frac{1}{2} \log \sin z - \frac{1}{2} \log \vartheta_1(z/\pi)$

2.33 $\sum_{1}^{\infty} [(-1)^n n^{-1}q^{2n}/(1-q^{2n})] \cos(2nz) = \frac{1}{2} \log(2Q_0) + \frac{1}{8} \log q + \frac{1}{2} \log \cos z - \frac{1}{2} \log \vartheta_2(z/\pi)$

2.34 $\sum_{1}^{\infty} [n^{-1}q^{2n}/(1-q^{2n})] \sin(2nz_1) \sin(2nz_2) = \frac{1}{4} \log \left\{\frac{\vartheta_0[(z_1+z_2)/\pi]}{\vartheta_0[(z_1-z_2)/\pi]}\right\}$

$|\operatorname{Im} z_1| + |\operatorname{Im} z_2| < \frac{1}{2}\pi \operatorname{Im}\tau$

2.35 $\sum_{1}^{\infty} [(-1)^n n^{-1}q^{2n}/(1-q^{2n})] \sin(2nz_1) \sin(2nz_2) = \frac{1}{4} \log \left(\frac{\vartheta_3[(z_1+z_2)/\pi]}{\vartheta_3[(z_1-z_2)/\pi]}\right)$

$|\operatorname{Im} z_1| + |\operatorname{Im} z_2| < \frac{1}{2}\pi \operatorname{Im}\tau$

2.36 $\sum_{1}^{\infty} [n^{-1}q^{2n}/(1-q^{2n})] \sin(2nz_1) \sin(2nz_2) = \frac{1}{4} \log \left(\frac{\vartheta_1[(z_1+z_2)/\pi]}{\vartheta_1[(z_1-z_2)/\pi]}\right) - \frac{1}{4} \log \left(\frac{\sin(z_1+z_2)}{\sin(z_1-z_2)}\right)$

$|\operatorname{Im} z_1| + |\operatorname{Im} z_2| < \pi \operatorname{Im}\tau$

2.37 $$\sum_{1}^{\infty} \frac{(-1)^n n^{-1} q^{2n}}{1-q^{2n}} \sin(2nz_1)\sin(2nz_2) = \frac{1}{4}\log\left(\frac{\vartheta_2[(z_1+z_2)/\pi]}{\vartheta_2[(z_1-z_2)/\pi]}\right) - \frac{1}{4}\log\left(\frac{\cos(z_1+z_2)}{\cos(z_1-z_2)}\right)$$

$|\operatorname{Im} z_1| + |\operatorname{Im} z_2| < \pi \operatorname{Im} \tau$

2.38 $$\sum_{0}^{\infty} \{[(2n)!]^2/(n!)^4\} 2^{-4n} \sin[(2n+\tfrac{1}{2})x] = \pi^{-1} K(\sin\tfrac{1}{2}x), \quad 0<x<\pi$$

2.39 $$\sum_{0}^{\infty} [(2n)!(2n+1)!/(n!)^3(n+1)!] 2^{-4n} \sin[(2n+\tfrac{3}{2})x]$$

$$= (2/\pi)[2E(\sin\tfrac{1}{2}x) - K(\sin\tfrac{1}{2}x)], \quad 0<x<\pi$$

2.40 $$\sum_{0}^{\infty} \{[(2n)!]^2/(n!)^4\} 2^{-4n} \cos[(2n+\tfrac{1}{2})x] = \pi^{-1} K(\cos\tfrac{1}{2}x), \quad 0<x<\pi$$

2.41 $$\sum_{0}^{\infty} [(2n)!(2n+1)!/(n!)^3(n+1)!] 2^{-2n} \cos[(2n+\tfrac{3}{2})x]$$

$$= (2/\pi)[K(\cos\tfrac{1}{2}x) - 2E(\cos\tfrac{1}{2}x)], \quad 0<x<\pi$$

2.42 $$\sin(n+1)x + [(n+1)/(2n+3)]\sin(n+3)x + \frac{1\cdot 3(n+1)(n+2)}{2!(2n+3)(2n+5)}\sin(n+5)x$$

$$+ \frac{1\cdot 3\cdot 5}{3!}\cdot\frac{(n+1)(n+2)(n+3)}{(2n+3)(2n+5)(2n+7)}\sin(n+7)x + \cdots$$

$$= \pi(2n+1)!(n!)^{-2} 2^{-2-2n} P_n(\cos x), \quad 0<x<\pi; \quad n=0,1,2,\ldots$$

2.43 $$\cos(n+1)x + \frac{n+1}{2n+3}\cos(n+3)x + \frac{1\cdot 3(n+1)(n+2)}{2!(2n+3)(2n+5)}\cos(n+5)x$$

$$+ \frac{1\cdot 3\cdot 5(n+1)(n+2)(n+3)}{3!(2n+3)(2n+5)(2n+7)}\cos(n+7)x + \cdots = (2n+1)!2^{-2n-1}(n!)^{-2} Q_n(\cos x)$$

$0<x<\pi; \quad n=0,1,2,\ldots$

2.44 $$\sum_{0}^{\infty} \frac{(2r+2n)!(l+r+n)!(l+1+n)!}{n!(r+n)!(2l+2+2n)!} \sin[(2n+l+r+1)x] = \pi 2^{-3-r-2l} (\sin x)^{-r} P_l^r(\cos x)$$

$0<x<\pi; \quad r, l=0,1,2,\ldots$

2.45 $$\sum_0^\infty \frac{(2r+2n)!(l+r+n)!(l+1+n)!}{n!(r+n)!(2l+2+2n)!}\cos[(2n+l+r+1)x]=2^{-2-r-2l}(\sin x)^{-r}Q_l^r(\cos x)$$

$$0<x<\pi;\quad r, l=0, 1, 2, \ldots$$

2.46 $$\sum_0^\infty [(\tfrac{1}{2}+\mu)_n(1+\nu+\mu)_n/n!(\nu+\tfrac{3}{2})_n]\sin[(2n+\nu+\mu+1)x]$$

$$=\pi^{\frac{1}{2}}2^{-1-\mu}[\Gamma(\tfrac{3}{2}+\nu)/\Gamma(\nu+\mu+1)](\sin x)^{-\mu}P_\nu^\mu(\cos x),\quad 0<x<\pi$$

2.47 $$\sum_0^\infty [(\tfrac{1}{2}+\mu)_n(1+\nu+\mu)_n/n!(\tfrac{3}{2}+\nu)_n]\cos[(2n+\nu+\mu+1)x]$$

$$=\pi^{-\frac{1}{2}}2^{-\mu}[\Gamma(\tfrac{3}{2}+\nu)/\Gamma(\nu+\mu+1)](\sin x)^{-\mu}Q_\nu^\mu(\cos x),\quad 0<x<\pi$$

2.48 $$\sum_1^\infty \log[n/(n+1)]\sin(2n+1)x=\sin x\psi(x/\pi)+\tfrac{1}{2}\pi\cos x+(\gamma+\log 2\pi)\sin x,\quad 0<x<\pi$$

2.49 $$\sum_1^\infty n^{-1}\log n\sin(nx)=\pi\log\Gamma(x/2\pi)+\tfrac{1}{2}\pi\log(\sin\tfrac{1}{2}x)$$

$$-\pi[1-(x/2\pi)]\log\pi-\tfrac{1}{2}\pi[1-(x/\pi)](\gamma+\log 2),\quad 0<x<2\pi$$

2.50 $$\sum_0^\infty \epsilon_n\{b^2+[n(\pi/a)]^2\}^{-\frac{1}{2}}\sin\{a[b^2+[n(\pi/a)]^2]^{\frac{1}{2}}\}\cos[n(\pi x/a)]=aJ_0[b(a^2-x^2)^{\frac{1}{2}}],$$

$$-a<x<a$$

2.51 $$\cos(\tfrac{1}{2}\pi z)\sum_1^\infty n^{z-1}\sin(2\pi n\alpha)+\sin(\tfrac{1}{2}\pi z)\sum_1^\infty n^{z-1}\cos(2\pi n\alpha)$$

$$=\sum_1^\infty n^{z-1}\sin(2\pi n\alpha+\tfrac{1}{2}\pi z)=\tfrac{1}{2}(2\pi)^{1-z}[\Gamma(1-z)]^{-1}\zeta(z,\alpha),\quad \mathrm{Re}z<1,\quad 0<\alpha<1$$

2.52 $$\cos(\tfrac{1}{2}\pi z)\sum_1^\infty (2n-1)^{z-1}\sin[(2n-1)\alpha\pi]+\sin(\tfrac{1}{2}\pi z)\sum_1^\infty (2n-1)^{z-1}\cos[(2n-1)\alpha\pi]$$

$$=\sum_1^\infty (2n-1)^{z-1}\sin[(2n-1)\alpha\pi+\tfrac{1}{2}\pi z]$$

$$=2^{-z-1}\pi^{1-z}[\Gamma(1-z)]^{-1}[\zeta(z,\tfrac{1}{2}\alpha)-\zeta(z,\tfrac{1}{2}+\tfrac{1}{2}\alpha)],\quad \mathrm{Re}z<1,\quad 0<\alpha\le 1$$

2.53 $$\sum_{0}^{\infty} \epsilon_n \cos(nx)\,[b^2+(nd)^2]^{-\frac{1}{2}} \sin\{a[b^2+(nd)^2]^{\frac{1}{2}}\} = (\pi/d) \sum_{m_1}^{m_2} J_0(b\{a^2-[(2\pi m+x)/d]^2\}^{\frac{1}{2}})$$

$$m_{\substack{1\\2}} = \mp[(ad \pm x)/2\pi]$$

If $(ad \pm x)/2\pi$ is an integer, then one half of the corresponding term in the sum is taken.

III

FOURIER SERIES WITH HIGHER FUNCTION COEFFICIENTS REPRESENTING ELEMENTARY FUNCTIONS

3.1 $\sum_{0}^{\infty} \{(-1)^n \epsilon_n \cos(2nx)/[\Gamma(1+\nu+n)\Gamma(1+\nu-n)]\} = [2^{2\nu}/\Gamma(1+2\nu)] \sin^{2\nu}x$

$\mathrm{Re}\nu > -\frac{1}{2}, \quad 0 < x < \pi$

3.2 $\sum_{0}^{\infty} \{(-1)^n \sin[(2n+1)x]/[\Gamma(\nu+\frac{3}{2}+n)\Gamma(\nu+\frac{1}{2}-n)]\} = [2^{2\nu-1}/\Gamma(1+2\nu)] \sin^{2\nu}x$

$\mathrm{Re}\nu > -\frac{1}{2}, \quad 0 < x < \pi$

3.3 $\sum_{0}^{\infty} \{\epsilon_n \cos(2nx)/[\Gamma(1+\nu+n)\Gamma(1+\nu-n)]\} = [2^{2\nu}/\Gamma(1+2\nu)] \cos^{2\nu}x$

$\mathrm{Re}\nu > -\frac{1}{2}, \quad -\frac{1}{2}\pi < x < \frac{1}{2}\pi$

3.4 $\sum_{0}^{\infty} \{\cos[(2n+1)x]/[\Gamma(\nu+\frac{3}{2}+n)\Gamma(\nu+\frac{1}{2}-n)]\} = [2^{2\nu-1}/\Gamma(1+2\nu)] \cos^{2\nu}x$

$\mathrm{Re}\nu > -\frac{1}{2}, \quad -\frac{1}{2}\pi < x < \frac{1}{2}\pi$

3.5 $\sum_{0}^{\infty} (-1)^n \epsilon_n J_{2n}(z) \cos(2nx) = \cos(z \cos x)$

3.6 $\sum_{0}^{\infty}(-1)^n J_{2n+1}(z)\cos[(2n+1)x]=\frac{1}{2}\sin(z\cos x)$

3.7 $\sum_{0}^{\infty}\epsilon_n J_{2n}(z)\cos(2nx)=\cos(z\sin x)$

3.8 $\sum_{0}^{\infty}J_{2n+1}(z)\sin[(2n+1)x]=\frac{1}{2}\sin(z\sin x)$

3.9 $\sum_{0}^{\infty}\epsilon_n J_{2n}(z)\cos(2nt)\cos(2nx)=\cos(z\sin x\cos t)\cos(z\cos x\cos t)$

3.10 $\sum_{0}^{\infty}J_{2n+1}(z)\sin[(2n+1)t]\cos[(2n+1)x]=\frac{1}{2}\cos(z\sin x\cos t)\sin(z\cos x\sin t)$

3.11 $\sum_{0}^{\infty}(-1)^n\epsilon_n J_{2n}(z)\cos(2nt)\cos(2nx)=\cos(z\cos x\cos t)\cos(z\sin x\sin t)$

3.12 $\sum_{0}^{\infty}(-1)^n J_{2n+1}(z)\sin[(2n+1)t]\sin[(2n+1)x]=\frac{1}{2}\cos(z\cos x\cos t)\sin(z\sin x\sin t)$

3.13 $\sum_{0}^{\infty}\epsilon_n I_n(z)\cos(nx)=\exp(z\cos x)$

3.14 $\sum_{0}^{\infty}(-1)^n\epsilon_n I_n(z)\cos(nx)=\exp(-z\cos x)$

3.15 $\sum_{0}^{\infty}\epsilon_n\exp[inm(\pi/2)]J_{nm}(z)\cos(nx)=m^{-1}\sum_{r=r_1}^{r_2}\exp\{iz\cos[(2\pi r+x)/m]\}$

$m=0,1,2,\ldots;\quad r_{\substack{1\\2}}=\mp[(m\pi\pm x)/2\pi]$

3.16 $\sum_{0}^{\infty}\epsilon_n I_{nm}(z)\cos(nx)=m^{-1}\sum_{r_1}^{r_2}\exp\{z\cos[(2\pi r+x)/m]\}$

$m=0,1,2,\ldots;\quad r_{\substack{1\\2}}=\mp[(m\pi\pm x)/2\pi]$

If $(m\pi \pm x)/2\pi$ is an integer or zero, one half of the corresponding term at the right side finite sum of 3.15 and 3.16 has to be taken. Formulas 3.5–3.16 hold also for arbitrary complex values of z, x.

$$\text{3.17}\quad \sum_1^\infty J_0(nx)\cos(nxt) = -\tfrac{1}{2} + \sum_{l=1}^{m} [x^2-(2\pi l+tx)^2]^{-\frac{1}{2}} + x^{-1}(1-t^2)^{-\frac{1}{2}} + \sum_{l=1}^{k} [x^2-(2\pi l-tx)^2]^{-\frac{1}{2}}$$

$$\text{3.18}\quad \sum_1^\infty J_0(nx)\sin(nxt) = (2\pi)^{-1}[\sum_{l=1}^{k} l^{-1} - \sum_{l=1}^{m} l^{-1}] + \sum_{l=m+1}^{\infty} \{[(2\pi l+tx)^2-x^2]^{-\frac{1}{2}} - (2\pi l)^{-1}\}$$
$$- \sum_{l=k+1}^{\infty} \{[(2\pi l-tx)^2-x^2]^{-\frac{1}{2}} - (2\pi l)^{-1}\}$$

$$\text{3.19}\quad \sum_1^\infty Y_0(nx)\cos(nxt) = -\pi^{-1}[\gamma+\log(x/4\pi)] + (2\pi)^{-1}[\sum_{l=1}^{m} l^{-1} + \sum_{l=1}^{k} l^{-1}]$$
$$- \sum_{l=m+1}^{\infty} \{[(2\pi l+tx)^2-x^2]^{-\frac{1}{2}} - (2\pi l)^{-1}\} - \sum_{l=k+1}^{\infty} \{[(2\pi l-tx)^2-x^2]^{-\frac{1}{2}} - (2\pi l)^{-1}\}$$

$$\text{3.20}\quad \sum_1^\infty J_0(nx)\cos(nxt) = -\tfrac{1}{2} + \sum_{l=m+1}^{k} [x^2-(2\pi l-tx)^2]^{-\frac{1}{2}}$$

$$\text{3.21}\quad \sum_1^\infty J_0(nx)\sin(nxt) = \sum_{l=0}^{m} [(2\pi l-tx)^2-x^2]^{-\frac{1}{2}} + (2\pi)^{-1}\sum_{l=1}^{k} l^{-1}$$
$$+ \sum_{l=1}^{\infty} \{[(2\pi l+tx)^2-x^2]^{-\frac{1}{2}} - (2\pi l)^{-1}\} - \sum_{l=k+1}^{\infty} \{[(2\pi l-tx)^2-x^2]^{-\frac{1}{2}} - (2\pi l)^{-1}\}$$

$$\text{3.22}\quad \sum_1^\infty Y_0(nx)\cos(nxt) = -\pi^{-1}[\gamma+\log(x/4\pi)] - \sum_{l=0}^{m} [(2\pi l-tx)^2-x^2]^{-\frac{1}{2}}$$
$$+(2\pi)^{-1}\sum_{l=1}^{k} l^{-1} - \sum_{l=1}^{\infty} \{[(2\pi l+tx)^2-x^2]^{-\frac{1}{2}} - (2\pi l)^{-1}\} - \sum_{l=1}^{\infty} \{[(2\pi l-tx)^2-x^2]^{-\frac{1}{2}} - (2\pi l)^{-1}\}$$

$$\text{3.23}\quad \sum_1^\infty (-1)^n J_0(nx)\cos(nxt) = -\tfrac{1}{2} + \sum_{l=1}^{m} [x^2-(2l\pi-\pi+tx)^2]^{-\frac{1}{2}} + \sum_{l=1}^{k} [x^2-(2l\pi-\pi-tx)^2]^{-\frac{1}{2}}$$

$$\text{3.24}\quad \sum_1^\infty (-1)^n J_0(nx)\sin(nxt) = (2\pi)^{-1}[\sum_{l=1}^{k} l^{-1} - \sum_{l=1}^{m} l^{-1}]$$
$$+ \sum_{l=m+1}^{\infty} \{[(2l\pi-\pi+tx)^2-x^2]^{-\frac{1}{2}} - (2\pi l)^{-1}\} - \sum_{l=k+1}^{\infty} \{[(2l\pi-\pi-tx)^2-x^2]^{-\frac{1}{2}} - (2\pi l)^{-1}\}$$

3.25 $$\sum_1^\infty (-1)^n Y_0(nxt)\cos(nxt) = -\pi^{-1}[\gamma+\log(x/4\pi)] + (2\pi)^{-1}[\sum_{l=1}^m l^{-1} + \sum_{l=1}^k l^{-1}]$$
$$- \sum_{l=m+1}^\infty \{[(2l\pi-\pi+tx)^2-x^2]^{-\frac12} - (2\pi l)^{-1}\} - \sum_{l=k+1}^\infty \{[(2l\pi-\pi-tx)^2-x^2]^{-\frac12} - (2\pi l)^{-1}\}$$

3.26 $$\sum_1^\infty (-1)^n J_0(nx)\cos(nxt) = -\tfrac12 + \sum_{l=m+1}^k [x^2-(2l\pi-\pi-tx)^2]^{-\frac12}$$

3.27 $$\sum_1^\infty (-1)^n J_0(nx)\sin(nxt) = (2\pi)^{-1}\sum_{l=1}^k l^{-1} + \sum_{l=1}^m [(2l\pi-\pi-tx)^2-x^2]^{-\frac12}$$
$$+ \sum_{l=1}^\infty \{[(2l\pi-\pi+tx)^2-x^2]^{-\frac12} - (2\pi l)^{-1}\} - \sum_{l=k+1}^\infty \{[(2l\pi-\pi-tx)^2-x^2]^{-\frac12} - (2\pi l)^{-1}\}$$

3.28 $$\sum_1^\infty (-1)^n Y_0(nx)\cos(nxt)$$
$$= -\pi^{-1}[\gamma+\log(x/4\pi)] + (2\pi)^{-1}\sum_{l=1}^k l^{-1} - \sum_{l=1}^m [(2l\pi-\pi-tx)^2-x^2]^{-\frac12}$$
$$- \sum_{l=1}^\infty \{[(2l\pi-\pi+tx)^2-x^2]^{-\frac12} - (2\pi l)^{-1}\} - \sum_{l=k+1}^\infty \{[(2l\pi-\pi-tx)^2-x^2]^{-\frac12} - (2\pi l)^{-1}\}$$

3.29 $$\sum_1^\infty K_0(nx)\cos(nxt) = \tfrac12[\gamma+\log(x/4\pi)] + (\pi/2x)(1+t^2)^{-\frac12}$$
$$+ \tfrac12\pi\sum_{l=1}^\infty \{[x^2+(2l\pi-tx)^2]^{-\frac12} - (2l\pi)^{-1}\} + \tfrac12\pi\sum_{l=1}^\infty \{[x^2+(2l\pi+tx)^2]^{-\frac12} - (2l\pi)^{-1}\}$$

3.30 $$\sum_1^\infty (-1)^n K_0(nx)\cos(nxt) = \tfrac12[\gamma+\log(x/4\pi)]$$
$$+ \tfrac12\pi\sum_{l=1}^\infty \{[x^2+(2l\pi-\pi-tx)^2]^{-\frac12} - (2l\pi)^{-1}\} + \tfrac12\pi\sum_{l=1}^\infty \{[x^2+(2l\pi-\pi+xt)^2]^{-\frac12} - (2l\pi)^{-1}\}$$

In 3.11–3.19: $0 \le t < 1, \quad x > 0, \quad m, k = 0, 1, 2, \ldots$

$$2\pi m < x(1-t) < 2(m+1)\pi; \qquad 2k\pi < x(1+t) < 2(k+1)\pi$$

In 3.20–3.22: $t > 1, \quad x > 0, \quad m, k = 0, 1, 2, \ldots$

$$2m\pi < x(t-1) < 2(m+1)\pi; \qquad 2k\pi < x(t+1) < 2(k+1)\pi$$

In 3.23–3.25: $0 < t \le 1, \quad x > 0, \quad m, k = 1, 2, 3, \ldots$

$$(2m-1)\pi < x(1-t) < (2m+1)\pi; \qquad (2k-1)\pi < x(1+t) < (2k+1)\pi$$

In 3.26–3.28: $t>1, \quad x>0, \quad m, k=1, 2, 3, \ldots$

$$(2m-1)\pi<x(t-1)<(2m+1)\pi; \qquad (2k-1)\pi<x(1+t)<(2k+1)\pi$$

3.31 $\sum_{0}^{\infty} \epsilon_n n^{-\nu} J_\nu(n\pi) \cos[n\pi(x/a)]=[2^{1-\nu}/\Gamma(\frac{1}{2}+\nu)]\pi^{\nu-\frac{1}{2}}a^{1-2\nu}(a^2-x^2)^{\nu-\frac{1}{2}}$

$-a<x<a, \quad \mathrm{Re}\,\nu>-\frac{1}{2}$

for $\nu=1$, Fourier series of a circle of radius a

3.32 $\sum_{0}^{\infty} \epsilon_n(-1)^n n^{-\nu} J_\nu(n\pi) \cos[2n\pi(x/a)]=[2^\nu/\Gamma(\frac{1}{2}+\nu)]\pi^{\nu-\frac{1}{2}}a^{1-2\nu}(ax-x^2)^{\nu-\frac{1}{2}}$

$\mathrm{Re}\,\nu>-\frac{1}{2}, \quad -a<x<a$

3.33 $\sum_{1}^{\infty} (-1)^n(2n+1)^{-\nu} J_\nu[(n+\frac{1}{2})\pi] \sin[(2n+1)(x/a)]=[a^{1-2\nu}/2\Gamma(\frac{1}{2}+\nu)]\pi^{\nu-\frac{1}{2}}(ax-x^2)^{\nu-\frac{1}{2}}$

$\mathrm{Re}\,\nu>-\frac{1}{2}, \quad 0<x<a$

3.34 $\sum_{1}^{\infty} n^{-\nu} J_{\nu+1}(n\pi) \sin[n\pi(x/a)]=[a^{-2\nu}/2\Gamma(\frac{1}{2}+\nu)](\frac{1}{2}\pi)^{\nu-\frac{1}{2}}x(a^2-x^2)^{\nu-\frac{1}{2}}$

$\mathrm{Re}\,\nu>-\frac{1}{2}, \quad 0<x<a$

3.35 $\sum_{0}^{\infty} (-1)^n J_0\{\frac{1}{2}a[b^2+(2n+1)^2(\pi^2/a^2)]^{\frac{1}{2}}\} \sin[(2n+1)\pi(x/a)]$

$=(a/2\pi)(ax-x^2)^{-\frac{1}{2}}\cos[b(ax-x^2)^{\frac{1}{2}}], \quad 0<x<a$

3.36 $\sum_{0}^{\infty} (-1)^n \epsilon_n J_0(\frac{1}{2}a\{b^2+[2n(\pi/a)]^2\}^{\frac{1}{2}}) \cos[2n\pi(x/a)]$

$=(a/\pi)(ax-x^2)^{-\frac{1}{2}}\cos[b(ax-x^2)^{\frac{1}{2}}], \quad -a<x<a$

3.37 $\sum_{0}^{\infty} (-1)^n J_0\{\frac{1}{2}a[(2n+1)^2(\pi^2/a^2)-b^2]^{\frac{1}{2}}\} \sin[(2n+1)\pi(x/a)]$

$=(a/2\pi)(ax-x^2)^{-\frac{1}{2}}\cosh[b(ax-x^2)^{\frac{1}{2}}], \quad 0<x<a$

3.38 $\sum_{0}^{\infty} (-1)^n \epsilon_n J_0(\frac{1}{2}a\{[2n(\pi/a)]^2-b^2\}^{\frac{1}{2}}) \cos[2\pi n(x/a)]$

$=(a/\pi)(ax-x^2)^{-\frac{1}{2}}\cosh[b(ax-x^2)^{\frac{1}{2}}], \quad -a<x<a$

3.39 $\sum_{0}^{\infty} \epsilon_n n^{\frac{1}{2}} J_\nu(\tfrac{1}{2}n\pi) J_{-\nu-\frac{1}{2}}(\tfrac{1}{2}n\pi) \cos[n\pi(x/a)]$

$$=\pi^{-2}(2a)^{\frac{3}{2}}(a^2x-x^3)^{-\frac{1}{2}} \cos[(2\nu+\tfrac{1}{2}) \arccos(x/a)], \quad 0<x<a$$

3.40 $\sum_{1}^{\infty} n^{\frac{1}{2}} J_{\nu+\frac{1}{2}}(\tfrac{1}{2}n\pi) J_{-\nu}(\tfrac{1}{2}n\pi) \sin[n\pi(x/a)]$

$$=\tfrac{1}{2}\pi^{-2}(2a)^{\frac{3}{2}}(a^2x-x^3)^{-\frac{1}{2}} \cos[(2\nu+\tfrac{1}{2}) \arccos(x/a)], \quad 0<x<a$$

3.41 $\sum_{0}^{\infty} n^{-\nu} \mathbf{H}_\nu(n\pi) \sin[n\pi(x/a)] = [2^{-\nu}/\Gamma(\tfrac{1}{2}+\nu)] a^{1-2\nu} \pi^{\nu-\frac{1}{2}} (a^2-x^2)^{\nu-\frac{1}{2}}$

$\mathrm{Re}\,\nu > -\tfrac{1}{2}, \quad 0<x<a$

for $\nu=1$, Fourier series of circle of radius a

3.42 $\sum_{0}^{\infty} \epsilon_n (nd)^{-\nu} J_\nu(and) \cos(nx) = [\pi^{\frac{1}{2}} 2^{1-\nu} a^{-\nu}/d\Gamma(\tfrac{1}{2}+\nu)] \sum_{m_1}^{m_2} \{a^2-[(2\pi m+x)/d]^2\}^{\nu-\frac{1}{2}}$

$\mathrm{Re}\,\nu > -\tfrac{1}{2}, \quad \mathrm{Re}\,\nu > \tfrac{1}{2} \text{ if } x=\pm ad$

3.43 $\sum_{0}^{\infty} \epsilon_n \cos(\tfrac{1}{2}and)(nd)^{-\nu} J_\nu(\tfrac{1}{2}and) \cos(nx)$

$$=[\pi^{\frac{1}{2}} a^{-\nu}/d\Gamma(\tfrac{1}{2}+\nu)] \sum_{m_1}^{m_2} |(2\pi m+x)/d|^{\nu-\frac{1}{2}} [a-|(2\pi m+x)/d|]^{\nu-\frac{1}{2}}$$

$\mathrm{Re}\,\nu > -\tfrac{1}{2}, \quad \mathrm{Re}\,\nu > \tfrac{1}{2} \text{ if } x=\pm ad, \text{ or } x=0$

3.44 $\sum_{0}^{\infty} \epsilon_n J_0\{a[b^2+(nd)^2]^{\frac{1}{2}}\} \cos(nx) = (2/d) \sum_{m_1}^{m_2} \{a^2-[(2\pi m+x)/d]^2\}^{-\frac{1}{2}}$

$$\cdot \cos(b\{a^2-[(2\pi m+x)/d]^2\}^{\frac{1}{2}}), \quad x\neq \pm ad$$

3.45 $\sum_{0}^{\infty} \epsilon_n J_0\{a[(nd)^2-b^2]^{\frac{1}{2}}\} \cos(nx) = (2/d) \sum_{m_1}^{m_2} \{a^2-[(2\pi m+x)/d]^2\}^{-\frac{1}{2}}$

$$\cdot \cosh(b\{a^2-[(2\pi m+x)/d]^2\}^{\frac{1}{2}}), \quad x\neq \pm ad$$

3.46 $\sum_{0}^{\infty} \epsilon_n \cos(\tfrac{1}{2}and) J_0\{\tfrac{1}{2}a[b^2+(nd)^2]^{\frac{1}{2}}\} \cos(nx) = d^{-\frac{1}{2}} \sum_{m_1}^{m_2} |(2\pi m+x)/d|^{-\frac{1}{2}}$

$$\cdot [a-|(2\pi m+x)/d|]^{-\frac{1}{2}} \cos\{b|(2\pi m+x)/d|^{\frac{1}{2}}[a-|(2\pi m+x)/d|^{\frac{1}{2}}]^{\frac{1}{2}}\}, \quad x\neq 0, \pm ad$$

3.47 $\sum_{0}^{\infty} \epsilon_n (nd)^{\frac{1}{2}} J_\nu(\tfrac{1}{2}and) J_{-\nu-\frac{1}{2}}(\tfrac{1}{2}and) \cos(nx) = 2(\tfrac{1}{2}\pi nd)^{-\frac{1}{2}} \sum_{m_1}^{m_2} |\, 2\pi m + x \,|^{-\frac{1}{2}}$

$\cdot \{a^2 - [(2\pi m + x)/d]^2\}^{-\frac{1}{2}} \cos\{(2\nu + \tfrac{1}{2}) \arccos[(2\pi m + x)/ad]\}, \quad x \neq 0, \pm ad$

3.48 $\sum_{0}^{\infty} \epsilon_n [\mathbf{J}_\nu(a+nd) + \mathbf{J}_{-\nu}(a+nd)] \cos(nx) = 4d^{-1} \cos(\tfrac{1}{2}\pi\nu) \sum_{m_1}^{m_2} \{a^2 - [(2\pi m + x)/d]^2\}^{-\frac{1}{2}}$

$\cdot \cos\{\nu \arccos[(2\pi m + x)/ad]\}, \quad x \neq \pm ad$

3.49 $\sum_{0}^{\infty} \epsilon_n [\mathbf{J}_{and/\pi}(z) + \mathbf{J}_{-and/\pi}(z)] \cos(nx) = -2\pi (ad)^{-1} \sum_{m_1}^{m_2} \sin\{z \sin[(\pi/ad)(2\pi m + x)]\}$

3.50 $\sum_{0}^{\infty} \epsilon_n [\cos(and)]^{-1} [\mathbf{J}_{2and/\pi}(z) + \mathbf{J}_{-2and/\pi}(z)] \cos(nx)$

$= 2\pi (ad)^{-1} \sum_{m_1}^{m_2} \cos\{z \cos[(\pi/2ad)(2\pi m + x)]\}$

3.51 $\sum_{0}^{\infty} \epsilon_n [\cos(and)]^{-1} [E_{2and/\pi}(z) + E_{-2and/\pi}(z)] \cos(nx)$

$= -2\pi (ad)^{-1} \sum_{m_1}^{m_2} \sin\{z \cos[(\pi/2ad)(2\pi m + x)]\}$

3.52 $\sum_{0}^{\infty} \epsilon_n [E_{and/\pi}(z) + E_{-and/\pi}(z)] \cos(nx) = -2\pi (ad)^{-1} \sum_{m_1}^{m_2} \sin\{z \sin[(\pi/ad)(2\pi m + x)]\}$

In Formulas 3.42–3.52 $m_1 = -[(ad+x)/2\pi]$; $m_2 = [(ad-x)/2\pi]$. If $(ad \pm x)/2\pi$ is an integer or zero, one half of the corresponding term in the right side sum has to be taken.

3.53 $\sum_{1}^{\infty} (\tfrac{1}{2}nz)^{-\nu} \mathbf{H}_\nu(nz) \sin(nx) = 0, \qquad 0 < z < x$

$= [\pi^{\frac{1}{2}}/z\Gamma(\tfrac{1}{2}+\nu)][1 - (x^2/z^2)]^{\nu-\frac{1}{2}}, \quad x < z < \pi$

3.54 $\sum_{0}^{\infty} P_n(\cos\vartheta) \sin[(n+\tfrac{1}{2})x] = 0, \qquad x < \vartheta$

$= -2^{-\frac{1}{2}}(\cos\vartheta - \cos x)^{-\frac{1}{2}}, \quad 0 < \vartheta < x < \pi$

$0 < \vartheta < \pi$

3.55 $\sum_0^\infty P_n(\cos\vartheta)\cos[(n+\tfrac{1}{2})x]=2^{-\frac{1}{2}}(\cos x-\cos\vartheta)^{-\frac{1}{2}}, \quad -\vartheta<x<\vartheta<\pi$

$=0, \quad \vartheta<x<2\pi-\vartheta$

$=-2^{-\frac{1}{2}}(\cos x-\cos\vartheta)^{-\frac{1}{2}}, \quad 2\pi-\vartheta<x<2\pi+\vartheta$

$=0, \quad 2\pi+\vartheta<x<4\pi-\vartheta$

$=2^{-\frac{1}{2}}(\cos x-\cos\vartheta)^{-\frac{1}{2}}, \quad 4\pi-\vartheta<x<4\pi+\vartheta$

$0<\vartheta<\pi$

3.56 $\sum_0^\infty \epsilon_n P_{n-\frac{1}{2}}(\cos\vartheta)\cos(nx)=2^{\frac{1}{2}}(\cos x-\cos\vartheta)^{-\frac{1}{2}}, \quad -\vartheta<x<\vartheta<\pi$

$=0, \quad \vartheta<x<2\pi-\vartheta$

$=2^{\frac{1}{2}}(\cos x-\cos\vartheta)^{-\frac{1}{2}}, \quad 2\pi-\vartheta<x<2\pi+\vartheta$

$0<\vartheta<\pi$

3.57 $\sum_0^\infty (-1)^n\epsilon_n P_{n-\frac{1}{2}}(-\cos\vartheta)\cos(nx)=0, \quad 0<x<\vartheta$

$=2^{\frac{1}{2}}(\cos\vartheta-\cos x)^{-\frac{1}{2}}, \quad \vartheta<x<\pi$

3.58 $\sum_0^\infty P_n(\cos\vartheta)\sin(nx)=-\sin(\tfrac{1}{2}x)(2\cos x-2\cos\vartheta)^{-\frac{1}{2}}, \quad x<\vartheta$

$=-\cos(\tfrac{1}{2}x)(2\cos\vartheta-2\cos x)^{-\frac{1}{2}}, \quad x>\vartheta$

3.59 $\sum_0^\infty P_n(\cos\vartheta)\cos(nx)=\cos(\tfrac{1}{2}x)(2\cos x-2\cos\vartheta)^{-\frac{1}{2}}, \quad x<\vartheta$

$=-\sin(\tfrac{1}{2}x)(2\cos\vartheta-2\cos x)^{-\frac{1}{2}}, \quad x>\vartheta$

3.60 $\sum_0^\infty \epsilon_n P_{-\frac{1}{2}+n\pi/\alpha}(\cos\vartheta)\cos(nx)=0, \quad (\pi/\alpha)\vartheta<x<2\pi-(\pi/\alpha)\vartheta$

$=2^{\frac{1}{2}}\pi^{-1}\alpha\{\cos[(\alpha/\pi)x]-\cos\vartheta\}^{-\frac{1}{2}}, \quad -(\pi/\alpha)\vartheta<x<(\pi/\alpha)\vartheta$

3.61 $\sum_0^\infty (-1)^n Q_n(z)\sin[(n+\tfrac{1}{2})x]=(2z+2\cos x)^{-\frac{1}{2}}\arctan\{[(1-\cos x)/(z+\cos x)]^{\frac{1}{2}}\}, \quad z>1$

3.62 $\sum_0^\infty Q_n(z)\cos[(n+\tfrac{1}{2})x]=(2z-2\cos x)^{-\frac{1}{2}}\arctan\{[(1+\cos x)/(z-\cos x)]^{\frac{1}{2}}\}, \quad z>1$

3.63 $\sum_0^\infty \epsilon_n Q_{n-\frac{1}{2}}(\cos\vartheta)\cos(nx)=\pi(2\cos\vartheta-2\cos x)^{-\frac{1}{2}}, \quad x>\vartheta$

$=0, \quad x<\vartheta$

3.64 $\sum_0^\infty \epsilon_n Q_{-\frac{1}{2}+\frac{1}{2}n}(z)\cos(nx)=[2/(z-\cos 2x)]^{\frac{1}{2}}(\frac{1}{2}\pi+\arctan\{[2/(z-\cos 2x)]^{\frac{1}{2}}\cos x\}),\quad z>1$

3.65 $\sum_0^\infty (-1)^n\epsilon_n Q_{-\frac{1}{2}+\frac{1}{2}n}(z)\cos(nx)=[2/(z-\cos 2x)]^{\frac{1}{2}}(\frac{1}{2}\pi-\arctan\{[2/(z-\cos 2x)]^{\frac{1}{2}}\cos x\})$

$z>1$

3.66 $\sum_0^\infty \epsilon_n(-1)^n Q_{n-\frac{1}{2}}(z)\cos(nx)=\pi(z+\cos x)^{-\frac{1}{2}},\quad z>1$

3.67 $\sum_0^\infty \epsilon_n P^{\mu}_{-\frac{1}{2}+n(\pi/\vartheta)}(\cos\vartheta)\cos[n\pi(x/a)]=(\frac{1}{2}\pi)^{-\frac{1}{2}}\vartheta[\Gamma(\frac{1}{2}-\mu)]^{-1}(\sin\vartheta)^{\mu}$

$\cdot\{\cos[(\vartheta/a)x]-\cos\vartheta\}^{-\mu-\frac{1}{2}},\quad \mathrm{Re}\mu<\frac{1}{2},\quad 0<\vartheta<\pi,\quad -a<x<a$

3.68 $\sum_0^\infty \epsilon_n P^{\mu}_{-\frac{1}{2}+n(\pi/\vartheta)}(\cosh\vartheta)\cos[n\pi(x/a)]=(\frac{1}{2}\pi)^{-\frac{1}{2}}\vartheta[\Gamma(\frac{1}{2}-\mu)]^{-1}(\sinh\vartheta)^{\mu}$

$\cdot\{\cos[(\vartheta/a)x]-\cos\vartheta\}^{-\mu-\frac{1}{2}},\quad \mathrm{Re}\mu<\frac{1}{2},\quad \vartheta>0,\quad -a<x<a$

3.69 $\sum_0^\infty \epsilon_n P^{\mu}_{-\frac{1}{2}+n}(\cos\vartheta)\cos(nx)=[(2\pi)^{\frac{1}{2}}/\Gamma(\frac{1}{2}-\mu)](\sin\vartheta)^{\mu}(\cos x-\cos\vartheta)^{-\mu-\frac{1}{2}},\quad 0<x<\vartheta$

$=0,\quad \vartheta<x<\pi$

$0<\vartheta<\pi,\quad \mathrm{Re}\mu<\frac{1}{2}$

3.70 $\sum_0^\infty \epsilon_n(-1)^n P^{\mu}_{n-\frac{1}{2}}(-\cos\vartheta)\cos(nx)$

$=[(2\pi)^{\frac{1}{2}}/\Gamma(\frac{1}{2}-\mu)](\sin\vartheta)^{\mu}(\cos x-\cos\vartheta)^{-\mu-\frac{1}{2}},\quad \vartheta<x<\pi$

$=0,\quad 0<x<\vartheta$

$0<\vartheta<\pi,\quad \mathrm{Re}\mu<\frac{1}{2}$

3.71 $\sum_0^\infty P_n{}^{\mu}(\cos\vartheta)\cos[(n+\frac{1}{2})x]=[(\frac{1}{2}\pi)^{\frac{1}{2}}/\Gamma(\frac{1}{2}-\mu)](\sin\vartheta)^{\mu}(\cos x-\cos\vartheta)^{-\mu-\frac{1}{2}},\quad 0<x<\vartheta$

$=0,\quad \vartheta<x<\pi$

$0<\vartheta<\pi,\quad \mathrm{Re}\mu<\frac{1}{2}$

3.72 $\sum_0^\infty \epsilon_n\{\Gamma[\nu+(a/\pi)nd]\Gamma[\nu-(a/\pi)nd]\}^{-1}\cos(nx)$

$=2^{2\nu-2}\pi[ad\Gamma(2\nu-1)]^{-1}\sum_{m_1}^{m_2}\{\cos[(\pi/2ad)(2\pi m+x)]\}^{2\nu-2}$

$\mathrm{Re}\nu>\frac{1}{2},\quad \mathrm{Re}\nu>1\ \text{ if }\ x=\pm ad$

3.73 $\sum_{0}^{\infty} \epsilon_n P^{\mu}_{-\frac{1}{2}+nd}(\cos a) \cos(nx) = [(2\pi)^{\frac{1}{2}}/d\Gamma(\frac{1}{2}-\mu)](\sin a)^{\mu}$

$\cdot \sum_{m_1}^{m_2} \{\cos[(2\pi m+x)/d] - \cos a\}^{-\mu-\frac{1}{2}}, \quad 0<a<\pi, \quad \operatorname{Re}\mu<\frac{1}{2}, \quad \operatorname{Re}\mu<-\frac{1}{2} \text{ if } x=\pm ad$

3.74 $\sum_{0}^{\infty} \epsilon_n P^{\mu}_{-\frac{1}{2}+ind}(\cosh a) \cos(nx) = [(2\pi)^{\frac{1}{2}}/d\Gamma(\frac{1}{2}-\mu)](\sinh a)^{\mu}$

$\cdot \sum_{m_1}^{m_2} \{\cosh a - \cosh[(2\pi m+x)/d]\}^{-\mu-\frac{1}{2}}, \quad \operatorname{Re}\mu<\frac{1}{2}, \quad \operatorname{Re}\mu<-\frac{1}{2} \text{ if } x=\pm ad$

3.75 $\sum_{0}^{\infty} \epsilon_n Q^{\mu}_{n-\frac{1}{2}}(z) \cos(nx) = \exp(i\pi\mu)(\frac{1}{2}\pi)^{\frac{1}{2}}\Gamma(\frac{1}{2}+\mu)(z^2-1)^{\frac{1}{2}\mu}(z-\cos x)^{-\mu-\frac{1}{2}}, \quad z>1, \quad \operatorname{Re}\mu>-\frac{1}{2}$

3.76 $\sum_{0}^{\infty} \epsilon_n Q^{\mu}_{-\frac{1}{2}+nm}(z) \cos(nx) = \exp(i\pi\mu)(2\pi)^{-\frac{1}{2}}\Gamma(\frac{1}{2}+\mu)(z^2-1)^{\frac{1}{2}\mu}$

$\cdot (\pi/m) \sum_{r_1}^{r_2} \{z - \cos[(2\pi r+x)/m]\}^{-\mu-\frac{1}{2}}$

$\operatorname{Re}\mu>-\frac{1}{2}, \quad m=1, 2, 3, \ldots, \quad r_1=-[(\pi m+x)/2\pi], \quad r_2=[(\pi m-x)/2\pi], \quad z>1$

If $(\pi m\pm x)/2\pi$ is an integer or zero, one half of the corresponding term at the right side has to be taken.

In 3.72–3.74, $m_1=-[(ad+x)/2\pi]$, $m_2=[(ad-x)/2\pi]$. If $(ad\pm x)/2\pi$ is an integer or zero, one half of the corresponding term at the right side has to be taken.

3.77 $\sum_{0}^{\infty} (-1)^n \epsilon_n \Gamma(\nu+2n+1) P_{\nu}^{-2n}(y) \cos(nx) = \Gamma(1+\nu)[y^2+(1-y^2)\cos^2 x]^{-\frac{1}{2}\nu-\frac{1}{2}}$

$\cdot \cos\{(\nu+1)\arctan[(y^{-2}-1)^{\frac{1}{2}}\cos x]\}, \quad 0<y<1, \quad \operatorname{Re}\nu>-1$

3.78 $\sum_{0}^{\infty} (-1)^n \Gamma(\nu+2n+2) P_{\nu}^{-2n-1}(y) \cos[(2n+1)x] = \frac{1}{2}\Gamma(\nu+1)[y^2+(1-y^2)\cos^2 x]^{-\frac{1}{2}\nu-\frac{1}{2}}$

$\cdot \sin\{(\nu+1)\arctan[(y^{-2}-1)^{\frac{1}{2}}\cos x\}, \quad 0<y<1, \quad \operatorname{Re}\nu>-1$

3.79 $\sum_{0}^{\infty} (-1)^n \epsilon_n Q^{\mu}_{n-\frac{1}{2}}(z)[\Gamma(\frac{1}{2}+n+\mu)\Gamma(\frac{1}{2}-n+\mu)]^{-1}\cos(2nx)$

$= [(\frac{1}{2}\pi)^{\frac{1}{2}}/\Gamma(\frac{1}{2}+\mu)](z^2-1)^{-\frac{1}{2}\mu}\exp(i\pi\mu)(1+z-2\cos^2 x)^{\mu-\frac{1}{2}}, \quad z>1$

3.80 $\sum_{0}^{\infty} [\epsilon_n(-i)^n/(n+m)!] P_l^n(\cos\vartheta)\cos(nx) = (l!)^{-1}(\cos\vartheta + i\sin\vartheta\cos x)^l, \quad l=1, 2, 3, \ldots$

3.81 $\sum_{0}^{\infty} [\epsilon_n(-i)^n/\Gamma(\nu+n+1)]P_\nu^n(\cos\vartheta) = [\Gamma(\nu+1)]^{-1}(\cos\vartheta + i\sin\vartheta\cos x)^\nu, \quad \mathrm{Re}\,\nu > -1$

3.82 $\sum_{0}^{\infty} [\epsilon_n/\Gamma(\nu+n+1)]P_\nu^n(\cosh z)\cos(nx) = [\Gamma(\nu+1)]^{-1}(\cosh z + \sinh z\cos x)^\nu, \quad \mathrm{Re}\,\nu > -1$

3.83 $\sum_{0}^{\infty} C_n^\nu(\cos\vartheta)\sin(nx) = -\sin(\nu x)(2\cos x - 2\cos\vartheta)^{-\nu}, \quad x < \vartheta$

$= -\sin[\nu(x+\pi)](2\cos\vartheta - 2\cos x)^{-\nu}, \quad x > \vartheta$

$0 < \vartheta < \pi, \quad \mathrm{Re}\,\nu < 1$

3.84 $\sum_{0}^{\infty} C_n^\nu(\cos\vartheta)\cos(nx) = \cos(\nu x)(2\cos x - 2\cos\vartheta)^{-\nu}, \quad x < \vartheta$

$= \cos[\nu(x+\pi)](2\cos\vartheta - 2\cos x)^{-\nu}, \quad x > \vartheta$

$0 < x < \vartheta, \quad \mathrm{Re}\,\nu < 1$

3.85 $\sum_{0}^{\infty} \epsilon_n A_n \cos(nx) = (1 - 2z\cos x + z^2)^{-\nu}$

$A_n = (z^n/n!)[\Gamma(n+\nu)/\Gamma(\nu)]{}_2F_1(\nu, n+\nu; n+1; z^2), \quad |z| < 1$

IV

FOURIER SERIES WITH HIGHER FUNCTION COEFFICIENTS REPRESENTING HIGHER FUNCTIONS

4.1 $\sum_{0}^{\infty} \epsilon_n[\Gamma(\frac{1}{4}+\frac{1}{2}n)/\Gamma(\frac{3}{4}+\frac{1}{2}n)]^2 \cos(nx) = 4K(\cos\frac{1}{2}x), \quad -\pi<x<\pi$

4.2 $\sum_{0}^{\infty} \epsilon_n(-1)^n[\Gamma(\frac{1}{4}+\frac{1}{2}n)/\Gamma(\frac{3}{4}+\frac{1}{2}n)]^2 \cos(nx) = 4K(\sin\frac{1}{2}x), \quad 0<x<2\pi$

4.3 $\sum_{0}^{\infty} \epsilon_n(-1)^n[\Gamma(\frac{1}{4}+\frac{1}{2}n)\Gamma(\frac{1}{4}-\frac{1}{2}n)/\Gamma(\frac{3}{4}+\frac{3}{2}n)\Gamma(\frac{3}{4}-\frac{3}{2}n)] \cos(nx) = 4K(\cos\frac{1}{2}x), \quad 0<x<\pi$

4.4 $\sum_{0}^{\infty} \epsilon_n[\Gamma(\frac{1}{4}+\frac{1}{2}n)\Gamma(\frac{1}{4}-\frac{1}{2}n)/\Gamma(\frac{3}{4}+\frac{1}{2}n)\Gamma(\frac{3}{4}-\frac{1}{2}n)] \cos(nx) = 4K(\sin\frac{1}{2}x), \quad 0<x<2\pi$

4.5 $\sum_{1}^{\infty} (-1)^n n^{-1}[\sin(n\pi s)\ \mathrm{Ci}(n\pi s) - \cos(n\pi s)\ \mathrm{si}(n\pi s)] \sin(nx)$
$= -\frac{1}{4}\{\psi[\frac{1}{2}+\frac{1}{2}s+(x/2\pi)] - \psi[\frac{1}{2}+\frac{1}{2}s-(x/2\pi)]\}, \quad -\pi\leq x\leq\pi, \quad s>0$

4.6 $$\sum_{0}^{\infty}(-1)^n\{\cos[(n+\tfrac{1}{2})\pi s]\,\mathrm{Ci}[(n+\tfrac{1}{2})\pi s]+\sin[(n+\tfrac{1}{2})\pi s]\,\mathrm{si}[(n+\tfrac{1}{2})\pi s]\}\,\sin[(n+\tfrac{1}{2})x]$$
$$=-\tfrac{1}{8}\{\psi[\tfrac{1}{4}+\tfrac{1}{4}s+(x/4\pi)]-\psi[\tfrac{1}{4}+\tfrac{1}{4}s-(x/4\pi)]-\psi[\tfrac{3}{4}+\tfrac{1}{4}s+(x/4\pi)]$$
$$+\psi[\tfrac{3}{4}+\tfrac{1}{4}s-(x/4\pi)]\},\quad s>0,\quad -\pi\le x\le\pi$$

4.7 $$\sum_{0}^{\infty}(-1)^n\{\sin[(n+\tfrac{1}{2})\pi s]\,\mathrm{Ci}[(n+\tfrac{1}{2})\pi s]-\cos[(n+\tfrac{1}{2})\pi s]\,\mathrm{si}[(n+\tfrac{1}{2})\pi s]\}\,\cos[(n+\tfrac{1}{2})x]$$
$$=\tfrac{1}{8}\{\psi[\tfrac{3}{4}+\tfrac{1}{4}s+(x/4\pi)]+\psi[\tfrac{3}{4}+\tfrac{1}{4}s-(x/4\pi)]-\psi[\tfrac{1}{4}+\tfrac{1}{4}s+(x/4\pi)]$$
$$-\psi[\tfrac{1}{4}+\tfrac{1}{4}s-(x/4\pi)]\},\quad -\pi\le x\le\pi,\quad s>0$$

4.8 $$\sum_{0}^{\infty}\epsilon_n J_n(z)\cos(nx)=\cos(z\sin x)+z\cos x\int_0^1\cos(tz\sin x)J_0(zt)\,dt$$

4.9 $$\sum_{1}^{\infty}J_n(z)\sin(nx)=\tfrac{1}{2}\sin(2\sin x)-\tfrac{1}{2}z\cos x\int_0^1\sin(tz\sin x)J_0(zt)\,dt$$

4.10 $$\sum_{0}^{\infty}(-1)^n J_{2n+\frac{1}{2}}(z)\cos[(2n+\tfrac{1}{2})x]=2^{-\frac{1}{2}}[\cos(z\cos x)C(2z\cos^2\tfrac{1}{2}x)+\sin(z\cos x)S(2z\cos^2\tfrac{1}{2}x)]$$

4.11 $$\sum_{0}^{\infty}(-1)^n J_{2n+\frac{3}{2}}(z)\cos[(2n+\tfrac{3}{2})x]=2^{-\frac{1}{2}}[\sin(z\cos x)C(2z\cos^2\tfrac{1}{2}x)-\cos(z\cos x)S(2z\cos^2\tfrac{1}{2}x)]$$

4.12 $$\sum_{1}^{\infty}I_n(z)\sin(nx)=\tfrac{1}{2}z\sin x\int_0^1\exp(-zt\cos x)I_0(zt)\,dt$$

4.13 $$\sum_{0}^{\infty}\epsilon_n J_0^2(\tfrac{1}{2}n\pi)\cos[n\pi(x/a)]=4\pi^{-2}K'(x/a),\quad 0<x<a$$

4.14 $$\sum_{0}^{\infty}\epsilon_n J_\nu(\tfrac{1}{2}n\pi)J_{-\nu}(\tfrac{1}{2}n\pi)\cos[n\pi(x/a)]=2\pi^{-1}P_{\nu-\frac{1}{2}}[2(x^2/a^2)-1],\quad -a<x<a$$

4.15 $$\sum_{0}^{\infty}\epsilon_n\{b^2+[n(\pi/a)]^2\}^{-\frac{1}{2}\nu}J_\nu(a\{b^2+[n(\pi/a)]^2\}^{\frac{1}{2}})\cos[n\pi(x/a)]$$
$$=(2/\pi a)^{\frac{1}{2}}(ab)^{\frac{1}{2}-\nu}(a^2-x^2)^{\frac{1}{2}\nu-\frac{1}{4}}J_{\nu-\frac{1}{2}}[b(a^2-x^2)^{\frac{1}{2}}],\quad \mathrm{Re}\,\nu>-\tfrac{1}{2},\quad -a<x<a$$

4.16 $$\sum_{0}^{\infty}\epsilon_n J_\nu(\tfrac{1}{2}a\{[b^2+n^2(\pi^2/a^2)]^{\frac{1}{2}}+n(\pi/a)\})J_\nu(\tfrac{1}{2}a\{[b^2+n^2(\pi^2/a^2)]^{\frac{1}{2}}-n(\pi/a)\})\cos[n\pi(x/a)]$$
$$=2a\pi^{-1}(a^2-x^2)^{-\frac{1}{2}}J_{2\nu}[b(a^2-x^2)^{\frac{1}{2}}],\quad \mathrm{Re}\,\nu>-\tfrac{1}{2},\quad -a<x<a$$

4.17 $\sum_0^\infty (-1)^n \epsilon_n J_\nu(\tfrac{1}{4}a\{[b^2+4n^2(\pi^2/a^2)]^{\frac{1}{2}}+2n(\pi/a)\})$

$\cdot J_\nu(\tfrac{1}{4}a\{[b^2+4n^2(\pi^2/a^2)]^{\frac{1}{2}}-2n(\pi/a)\}) \cos[2n\pi(x/a)]$

$= a\pi^{-1}(ax-x^2)^{-\frac{1}{2}} J_{2\nu}[b(a^2-x^2)^{\frac{1}{2}}], \quad \text{Re}\nu > -\tfrac{1}{2}, \quad -a<x<a$

4.18 $\sum_0^\infty (-1)^n J_\nu(\tfrac{1}{4}a\{[b^2+(2n+1)^2(\pi^2/a^2)]^{\frac{1}{2}}+(2n+1)(\pi/a)\})$

$\cdot J_\nu(\tfrac{1}{4}a\{[b^2+(2n+1)^2(\pi^2/a^2)]^{\frac{1}{2}}-(2n+1)(\pi/a)\}) \sin[(2n+1)\pi(x/a)]$

$= \tfrac{1}{2}a\pi^{-1}(ax-x^2)^{-\frac{1}{2}} J_{2\nu}[b(a^2-x^2)], \quad \text{Re}\nu > -\tfrac{1}{2}, \quad 0<x<a$

4.19 $\sum_0^\infty \epsilon_n I_{\frac{1}{2}n}(z) \cos(nx) = \exp(z\cos 2x)\{1+\text{Erf}[(2z)^{\frac{1}{2}}\cos x]\}$

$= \exp(z\cos 2x)\{2-\text{Erfc}[(2z)^{\frac{1}{2}}\cos x]\}$

4.20 $\sum_0^\infty (-1)^n \epsilon_n I_{\frac{1}{2}n}(z) \cos(nx) = \exp(z\cos 2x)\ \text{Erfc}[(2z)^{\frac{1}{2}}\cos x]$

$= \exp(z\cos 2x)\{1-\text{Erf}[(2z)^{\frac{1}{2}}\cos x]\}$

4.21 $\sum_0^\infty I_{n+\frac{1}{2}}(z) \cos[(2n+1)x] = \tfrac{1}{2}\exp(z\cos 2x)\ \text{Erf}[(2z)^{\frac{1}{2}}\cos x]$

4.22 $\sum_0^\infty (-1)^n I_{n+\frac{1}{2}}(z) \sin[(2n+1)x] = \tfrac{1}{2}\exp(-z\cos 2x)\ \text{Erf}[(2z)^{\frac{1}{2}}\sin x]$

4.23 $\sum_0^\infty (-1)^n I_{n+\frac{1}{2}}(z) \cos[(2n+1)x] = -\tfrac{1}{2}i\exp(-z\cos 2x)\ \text{Erf}[i(2z)^{\frac{1}{2}}\cos x]$

4.24 $\sum_0^\infty I_{n+\frac{1}{2}}(z) \sin[(2n+1)x] = -\tfrac{1}{2}i\exp(z\cos 2x)\ \text{Erf}[i(2z)^{\frac{1}{2}}\sin x]$

4.25 $\sum_0^\infty \epsilon_n J_{n-\frac{1}{4}}(z) J_{n+\frac{1}{4}}(z) \cos(2nx)$

$= 2(2\pi z\sin x)^{-\frac{1}{2}}[\cos(2z\sin x)\,C(2z\sin x)+\sin(2z\sin x)\,S(2z\sin x)]$

4.26 $\sum_0^\infty (-1)^n \epsilon_n J_{n-\frac{1}{4}}(z) J_{n+\frac{1}{4}}(z) \cos(2nx)$

$= 2(2\pi z\cos x)^{-\frac{1}{2}}[\cos(2z\cos x)\,C(2z\cos x)+\sin(2z\cos x)\,S(2z\cos x)]$

4.27 $$\sum_{0}^{\infty} (-1)^n J_{n+\frac{1}{4}}(z) J_{n+\frac{3}{4}}(z) \cos[(2n+1)x]$$
$$= (2\pi z \cos x)^{-\frac{1}{2}}[\sin(2z \cos x) C(2z \cos x) - \cos(2z \cos x) S(2z \cos x)]$$

4.28 $$\sum_{0}^{\infty} J_{n+\frac{1}{4}}(z) J_{n+\frac{3}{4}}(z) \sin[(2n+1)x]$$
$$= (2\pi z \sin x)^{-\frac{1}{2}}[\sin(2z \sin x) C(2z \sin x) - \cos(2z \sin x) S(2z \sin x)]$$

4.29 $$\sum_{0}^{\infty} \epsilon_n J_n^2(z) \cos(nx) = J_0(2z \sin\tfrac{1}{2}x)$$

4.30 $$\sum_{0}^{\infty} \epsilon_n J_n(z) Y_n(z) \cos(nx) = Y_0(2z \mid \sin\tfrac{1}{2}x \mid)$$

4.31 $$\sum_{0}^{\infty} \epsilon_n (-1)^n [I_n(z)]^2 \cos(nx) = I_0(2z \sin\tfrac{1}{2}x)$$

4.32 $$\sum_{0}^{\infty} \epsilon_n I_n(z) K_n(z) \cos(nx) = K_0(2z \mid \sin\tfrac{1}{2}x \mid)$$

4.33 $$\sum_{0}^{\infty} J_n(z) J_{n+1}(z) \sin[(2n+1)x] = \tfrac{1}{2} J_1(2z \sin x)$$

4.34 $$\sum_{0}^{\infty} (-1)^n J_n(z) J_{n+1}(z) \cos[(2n+1)x] = \tfrac{1}{2} J_1(2z \cos x)$$

4.35 $$\sum_{0}^{\infty} (-1)^n [J_{n+\frac{1}{2}}(z)]^2 \cos[(2n+1)x] = \tfrac{1}{2} \mathbf{H}_0(2z \cos x)$$

4.36 $$\sum_{0}^{\infty} [J_{n+\frac{1}{2}}(z)]^2 \sin[(2n+1)x] = -\tfrac{1}{2} \mathbf{H}_0(2z \sin x)$$

4.37 $$\sum_{0}^{\infty} (-1)^{n+1} J_{n+\frac{1}{2}}(z) Y_{n+\frac{1}{2}}(z) \cos[(2n+1)x] = \tfrac{1}{2} J_0(2z \cos x)$$

4.38 $$\sum_{0}^{\infty} J_{n+\frac{1}{2}}(z) Y_{n+\frac{1}{2}}(z) \sin[(2n+1)x] = -\tfrac{1}{2} J_0(2z \sin x)$$

4.39 $\sum_0^\infty \epsilon_n J_n(z_1) J_n(z_2) \cos(nx) = J_0[(z_1^2+z_2^2-2z_1z_2 \cos x)^{\frac{1}{2}}]$

4.40 $\sum_0^\infty \epsilon_n J_n(z_1) Y_n(z_2) \cos(nx) = Y_0[(z_1^2+z_2^2-2z_1z_2 \cos x)^{\frac{1}{2}}], \quad z_2 > z_1$

4.41 $\sum_0^\infty (-1)^n \epsilon_n I_n(z_1) I_n(z_2) \cos(nx) = I_0[(z_1^2+z_2^2-2z_1z_2 \cos x)^{\frac{1}{2}}]$

4.42 $\sum_0^\infty \epsilon_n I_n(z_1) K_n(z_2) \cos(nx) = K_0[(z_1^2+z_2^2-2z_1z_2 \cos x)^{\frac{1}{2}}], \quad z_2 > z_1$

4.43 $\sum_0^\infty (-1)^n \epsilon_n I_{nm}(z_1) I_{nm}(z_2) \cos(nx) = m^{-1} \sum_{r_1}^{r_2} I_0(\{z_1^2+z_2^2-2z_1z_2 \cos[(2\pi r+x)/m]\}^{\frac{1}{2}})$

4.44 $\sum_0^\infty \epsilon_n J_{nm}(z_1) J_{nm}(z_2) \cos(nx) = m^{-1} \sum_{r_1}^{r_2} J_0(\{z_1^2+z_2^2-2z_1z_2 \cos[(2\pi r+x)/m]\}^{\frac{1}{2}})$

4.45 $\sum_0^\infty \epsilon_n J_{nm}(z_1) Y_{nm}(z_2) \cos(nx) = m^{-1} \sum_{r_1}^{r_2} Y_0(\{z_1^2+z_2^2-2z_1z_2 \cos[(2\pi r+x)/m]\}^{\frac{1}{2}})$

4.46 $\sum_0^\infty \epsilon_n I_{nm}(z_1) K_{nm}(z_2) \cos(nx) = m^{-1} \sum_{r_1}^{r_2} K_0(\{z_1^2+z_2^2-2z_1z_2 \cos[(2\pi r+x)/m]\}^{\frac{1}{2}})$

In the last four formulas $m = 1, 2, 3, \ldots, r_1 = -[(m\pi+x)/2\pi], r_2 = [(m\pi-x)/2\pi]$. If $(m\pi \pm x)/2\pi$ is an integer or zero, one half of the corresponding term in the sum at the right side has to be taken.

4.47 $\sum_0^\infty J_{n+\frac{1}{2}}(az) J_{n+\frac{1}{2}}(bz) \cos[(n+\tfrac{1}{2})x] = \pi^{-1} \int_{t_1}^{t_2} [t^2-(a^2+b^2-2ab \cos x)]^{-\frac{1}{2}} \sin(tz)\, dt$

$t_1 = (a^2+b^2-2ab \cos x)^{\frac{1}{2}}, \quad t_2 = a+b$

4.48 $\sum_0^\infty (-1)^{n+1} J_{n+\frac{1}{2}}(az) Y_{n+\frac{1}{2}}(bz) \cos[(n+\tfrac{1}{2})x]$

$= \pi^{-1} \int_{t_1}^{t_2} [(a^2+b^2+2ab \cos x)-t^2]^{-\frac{1}{2}} \cos(tz)\, dt$

$t_1 = b-a, \quad t_2 = (a^2+b^2+2ab \cos x)^{\frac{1}{2}}; \quad b > a$

4.49 $$\sum_{0}^{\infty} (-1)^n \epsilon_n J_{2n}(z) K_{2n}(z) \cos(2nx) = \tfrac{1}{2}\{K_0[z(2i \cos x)^{\frac{1}{2}}] + K_0[z(-2i \cos x)^{\frac{1}{2}}]\}$$

4.50 $$\sum_{0}^{\infty} (-1)^n J_{2n+1}(z) K_{2n+1}(z) \cos[(2n+1)x] = \tfrac{1}{4} i\{K_0[z(2i \cos x)^{\frac{1}{2}}] - K_0[z(-2i \cos x)^{-\frac{1}{2}}]\}$$

4.51 $$\sum_{0}^{\infty} \epsilon_n J_{2n}(z) K_{2n}(z) \cos(2nx) = \tfrac{1}{2}\{K_0[z(2i \sin x)^{\frac{1}{2}}] + K_0[z(-2i \sin x)^{\frac{1}{2}}]\}$$

4.52 $$\sum_{0}^{\infty} J_{2n+1}(z) K_{2n+1}(z) \sin[(2n+1)x] = \tfrac{1}{4} i\{K_0[z(2i \sin x)^{\frac{1}{2}}] - K_0[z(-2i \sin x)^{\frac{1}{2}}]\}$$

4.53 $$\sum_{0}^{\infty} (-1)^n I_{n+\frac{1}{2}}(z) K_{n+\frac{1}{2}}(z) \cos[(2n+1)x] = \tfrac{1}{4}\pi[I_0(2z \cos x) - L_0(2z \cos x)]$$

4.54 $$\sum_{0}^{\infty} I_{n+\frac{1}{2}}(x) K_{n+\frac{1}{2}}(x) \sin[(2n+1)x] = \tfrac{1}{4}\pi[I_0(2z \sin x) - L_0(2z \sin x)]$$

4.55 $$\sum_{1}^{\infty} J_n(z) J_{n+2m}(z) \cos[2x(n+m)] = \tfrac{1}{2}(-1)^m J_{2m}(2z \sin x) - \tfrac{1}{2}(-1)^m \cdot \sum_{k=0}^{m} \epsilon_k (-1)^k J_{m-k}(z) J_{m+k}(z) \cos(2kx), \quad m = 0, 1, 2, \ldots$$

4.56 $$\sum_{1}^{\infty} (-1)^n J_n(z) J_{n+2m}(z) \cos[2x(n+m)] = \tfrac{1}{2} J_{2m}(2z \cos x) - \tfrac{1}{2} \sum_{k=0}^{m} \epsilon_k J_{m-k}(z) J_{m+k}(z) \cos(2kx), \quad m = 0, 1, 2, \ldots$$

4.57 $$\sum_{0}^{\infty} (-1)^n J_n(z) J_{n+2m+1}(z) \cos[x(2n+2m+1)] = \tfrac{1}{2} J_{2m+1}(2z \cos x) - \sum_{k=0}^{m-1} J_{m-k}(z) J_{m+k+1}(z) \cos[(2k+1)x]$$

$$m = 0, 1, 2, \ldots; \quad \sum_{k=0}^{m-1} (\ ) = 0 \quad \text{if} \quad m = 0$$

4.58 $\sum_{0}^{\infty} J_n(z)J_{n+2m+1}(z)\ \sin[x(2n+2m+1)]$

$$=\tfrac{1}{2}(-1)^m J_{2m+1}(2z\ \sin x)-(-1)^m\sum_{k=0}^{m-1}(-1)^k J_{m-k}(z)J_{m+k+1}(z)\ \sin[(2k+1)x]$$

$$m=0,1,2,\ldots;\quad \sum_{k=0}^{m-1}(\ )=0\quad\text{if}\quad m=0$$

4.59 $\sum_{0}^{\infty}(-1)^n\epsilon_n J_{n-\nu}(z)J_{n+\nu}(z)\ \cos(2nx)=[2\ \cos(\pi\nu)]^{-1}[\mathbf{J}_{2\nu}(2z\ \cos x)+\mathbf{J}_{-2\nu}(2z\ \cos x)]$

$$=[2\ \sin(\pi\nu)]^{-1}[E_{2\nu}(2z\ \cos x)-E_{-2\nu}(2z\ \cos x)]$$

4.60 $\sum_{0}^{\infty}(-1)^n J_{n+\frac{1}{2}+\nu}(z)J_{n+\frac{1}{2}-\nu}(z)\ \cos[(2n+1)x]$

$$=[4\ \sin(\pi\nu)]^{-1}[\mathbf{J}_{2\nu}(2z\ \cos x)-\mathbf{J}_{-2\nu}(2z\ \cos x)]$$

$$=-[4\ \cos(\pi\nu)]^{-1}[E_{2\nu}(2z\ \cos x)+E_{-2\nu}(2z\ \cos x)]$$

4.61 $\sum_{0}^{\infty}(-1)^n\epsilon_n[J_{n+\frac{1}{2}\nu}(z)Y_{n-\frac{1}{2}\nu}(z)+J_{n-\frac{1}{2}\nu}(z)Y_{n+\frac{1}{2}\nu}(z)]\cos(2nx)$

$$=2\ \sin(\tfrac{1}{2}\pi\nu)J_\nu(2z\ \cos x)+2\ \cos(\tfrac{1}{2}\pi\nu)Y_\nu(2z\ \cos x),\quad \operatorname{Re}\nu>-1$$

4.62 $\sum_{0}^{\infty}(-1)^{n+1}[J_{n+\frac{1}{2}+\frac{1}{2}\nu}(z)Y_{n+\frac{1}{2}-\frac{1}{2}\nu}(z)+J_{n+\frac{1}{2}-\frac{1}{2}\nu}(z)Y_{n+\frac{1}{2}+\frac{1}{2}\nu}(z)]\cos[(2n+1)x]$

$$=\cos(\tfrac{1}{2}\pi\nu)J_\nu(2z\ \cos x)-\sin(\tfrac{1}{2}\pi\nu)Y_\nu(2z\ \cos x),\quad \operatorname{Re}\nu>-1$$

4.63 $\sum_{0}^{\infty}(-1)^n\epsilon_n J_{\nu-n}(z)J_{\nu+n}(z)\ \cos(2nx)=J_{2\nu}(2z\,|\sin x|),\quad \operatorname{Re}\nu>-\tfrac{1}{2}$

4.64 $\sum_{0}^{\infty}\epsilon_n J_{\nu-n}(z)J_{\nu+n}(z)\ \cos(2nx)=J_{2\nu}(2z\ \cos x),\quad \operatorname{Re}\nu>-\tfrac{1}{2}$

4.65 $\sum_{0}^{\infty}(-1)^n J_{\nu-n-\frac{1}{2}}(z)J_{\nu+n+\frac{1}{2}}(z)\ \sin[(2n+1)x]=\tfrac{1}{2}J_{2\nu}(2z\,|\sin x|),\quad \operatorname{Re}\nu>-\tfrac{1}{2}$

4.66 $\sum_{0}^{\infty}J_{\nu-n-\frac{1}{2}}(z)J_{\nu+n+\frac{1}{2}}(z)\ \cos[(2n+1)x]=\tfrac{1}{2}J_{2\nu}(2z\ \cos x),\quad \operatorname{Re}\nu>-\tfrac{1}{2}$

4.67 $\sum_{0}^{\infty}(-1)^n\epsilon_n I_{\nu-n}(z)I_{\nu+n}(z)\ \cos(2nx)=I_{2\nu}(2z\,|\sin x|),\quad \operatorname{Re}\nu>-\tfrac{1}{2}$

4.68 $\sum_{0}^{\infty} \epsilon_n I_{\nu-n}(z) I_{\nu+n}(z) \cos(2nx) = I_{2\nu}(2z \cos x), \quad \mathrm{Re}\nu > -\frac{1}{2}$

4.69 $\sum_{0}^{\infty} (-1)^n I_{\nu-n-\frac{1}{2}}(z) I_{\nu+n+\frac{1}{2}}(z) \sin[(2n+1)x] = \frac{1}{2} I_{2\nu}(2z \sin x), \quad \mathrm{Re}\nu > -\frac{1}{2}$

4.70 $\sum_{0}^{\infty} I_{\nu-n-\frac{1}{2}}(z) I_{\nu+n+\frac{1}{2}}(z) \cos[(2n+1)z] = \frac{1}{2} I_{2\nu}(2z \cos x), \quad \mathrm{Re}\nu > -\frac{1}{2}$

4.71 $\sum_{0}^{\infty} (-1)^n \epsilon_n I_{n-\frac{1}{2}\nu}(z) K_{n+\frac{1}{2}\nu}(z) \cos(2nx)$

$= -[\pi/\sin(\pi\nu)][\frac{1}{2}\mathbf{J}_{-\nu}(2iz \cos x) + \frac{1}{2}\mathbf{J}_{-\nu}(-2iz \cos x) - \cos(\frac{1}{2}\pi\nu) I_{-\nu}(2z \cos x)], \quad \mathrm{Re}\nu < 1$

4.72 $\sum_{0}^{\infty} (-1)^n I_{n+\frac{1}{2}+\frac{1}{2}\nu}(z) K_{n+\frac{1}{2}-\frac{1}{2}\nu}(z) \cos[(2n+1)x] = [\pi/2 \sin(\pi\nu)][\frac{1}{2}i\mathbf{J}_{\nu}(2iz \cos x)$

$-\frac{1}{2}i\mathbf{J}_{\nu}(-2iz \cos x) + \sin(\frac{1}{2}\pi\nu) I_{\nu}(2z \cos x)], \quad \mathrm{Re}\nu > -1$

4.73 $\sum_{0}^{\infty} (-1)^n \epsilon_n [I_{n+\nu}(z) K_{n-\nu}(z) + I_{n-\nu}(z) K_{n+\nu}(z)] \cos(2nx) = 2 \cos(\pi\nu) K_{2\nu}(2z \cos x)$

$\mathrm{Re}\nu > -\frac{1}{2}$

4.74 $\sum_{0}^{\infty} (-1)^n [I_{n+\frac{1}{2}-\nu}(z) K_{n+\frac{1}{2}+\nu}(z) - I_{n+\frac{1}{2}+\nu}(z) K_{n+\frac{1}{2}-\nu}(z)] \cos[(2n+1)x]$

$= 2 \sin(\pi\nu) K_{2\nu}(2z \cos x), \quad \mathrm{Re}\nu > -\frac{1}{2}$

4.75 $\sum_{0}^{\infty} \epsilon_n (nd)^{-\nu} J_{\nu+2l}(nda) \cos(nx)$

$= [(-1)^l (\frac{1}{2}a)^{-\nu} (2l)!/d\Gamma(2l+2\nu)] \sum_{m_1}^{m_2} \{a^2 - [(2\pi m+x)/d]^2\}^{\nu-\frac{1}{2}} C_{2l}^{\nu}[(2\pi m+x)/ad]$

$l = 0, 1, 2, \ldots; \quad \mathrm{Re}\nu > -\frac{1}{2}; \quad \mathrm{Re}\nu > \frac{1}{2} \text{ if } x = \pm ad$

4.76 $\sum_{0}^{\infty} \epsilon_n [J_0(\frac{1}{2}and)]^2 \cos(nx) = (4/\pi ad) \sum_{m_1}^{m_2} K'[|(2\pi m+x)/ad|]$

4.77 $\sum_{0}^{\infty} \epsilon_n J_{\nu}(\frac{1}{2}and) J_{-\nu}(\frac{1}{2}and) \cos(nx) = (2/ad) \sum_{m_1}^{m_2} P_{\nu-\frac{1}{2}}[(2/a^2d^2)(2\pi m+x)^2 - 1]$

4.78 $$\sum_{0}^{\infty} \epsilon_n [b^2+(nd)^2]^{-\frac{1}{2}\nu} J_\nu\{a[b^2+(nd)^2]^{\frac{1}{2}}\} \cos(nx)$$
$$= (2\pi b^{\frac{1}{2}}/d)(ab)^{-\nu} \sum_{m_1}^{m_2} \{a^2-[(2\pi m+x)/d]^2\}^{\frac{1}{2}\nu-\frac{1}{4}} J_{\nu-\frac{1}{2}}(b\{a^2-[(2\pi m+x)/d]^2\}^{\frac{1}{2}})$$
$$\mathrm{Re}\nu > -\tfrac{1}{2};\quad \mathrm{Re}\nu > \tfrac{1}{2} \quad \text{if} \quad x = \pm ad$$

4.79 $$\sum_{0}^{\infty} \epsilon_n T_{2l}\{nd/[b^2+(nd)^2]^{\frac{1}{2}}\} J_{2l}\{a[b^2+(nd)^2]^{\frac{1}{2}}\} \cos(nx)$$
$$= (-1)^l (2/d) \sum_{m_1}^{m_2} \{a^2-[(2\pi m+x)/d]^2\}^{-\frac{1}{2}} \cos(b\{a^2-[(2\pi m+x)/d]^2\}^{\frac{1}{2}})$$
$$\cdot T_{2l}[(2\pi m+x)/ad], \quad l=0,1,2,\ldots; \quad x \neq ad$$

4.80 $$\sum_{0}^{\infty} \epsilon_n T_{2l}\{b[b^2+(nd)^2]^{-\frac{1}{2}}\} J_{2l}\{a[b^2+(nd)^2]^{\frac{1}{2}}\} \cos(nx)$$
$$= (-1)^l (2/d) \sum_{m_1}^{m_2} \{a^2-[(2\pi m+x)/d]^2\}^{-\frac{1}{2}} \cos(b\{a^2-[(2\pi m+x)/d]^2\}^{\frac{1}{2}})$$
$$\cdot T_{2l}(\{1-[(2\pi m+x)/ad]^2\}^{\frac{1}{2}}), \quad l=0,1,2,\ldots; \quad x \neq ad$$

4.81 $$\sum_{0}^{\infty} \epsilon_n T_{2l+1}\{b[b^2+(nd)^2]^{-\frac{1}{2}}\} J_{2l+1}\{a[b^2+(nd)^2]^{\frac{1}{2}}\} \cos(nx)$$
$$= (-1)^l (2/d) \sum_{m_1}^{m_2} \{a^2-[(2\pi m+x)/d]^2\}^{-\frac{1}{2}} \sin(b\{a^2-[(2\pi m+x)/d]^2\}^{\frac{1}{2}})$$
$$\cdot T_{2l+1}(\{1-[(2\pi m+x)/ad]^2\}^{\frac{1}{2}}), \quad l=0,1,2,\ldots$$

4.82 $$\sum_{0}^{\infty} \epsilon_n [b^2+(nd)^2]^{-\frac{1}{2}} U_{2l}\{nd[b^2+(nd)^2]^{-\frac{1}{2}}\} J_{2l+1}\{a[b^2+(nd)^2]^{\frac{1}{2}}\} \cos(nx)$$
$$= (-1)^l (2abd)^{-1} \sum_{m_1}^{m_2} \sin(b\{a^2-[(2\pi m+x)/d]^2\}^{\frac{1}{2}}) U_{2l}[(2\pi m+x)/ad], \quad l=0,1,2,\ldots$$

4.83 $$\sum_{0}^{\infty} \epsilon_n J_\nu(\tfrac{1}{2}a\{[b^2+(nd)^2]^{\frac{1}{2}}+nd\}) J_\nu(\tfrac{1}{2}a\{[b^2+(nd)^2]^{\frac{1}{2}}-nd\}) \cos(nx)$$
$$= (2/d) \sum_{m_1}^{m_2} \{a^2-[(2\pi m+x)/d]^2\}^{-\frac{1}{2}} J_{2\nu}(b\{a^2-[(2\pi m+x)/d]^2\}^{\frac{1}{2}})$$
$$\mathrm{Re}\nu > -\tfrac{1}{2};\quad \mathrm{Re}\nu > -\tfrac{1}{2} \quad \text{if} \quad x = \pm ad$$

4.84 $\displaystyle\sum_{0}^{\infty} \epsilon_n \cos(\tfrac{1}{2}and) J_\nu(\tfrac{1}{4}a\{[b^2+(nd)^2]^{\frac12}+nd\}) J_\nu(\tfrac{1}{4}a\{[b^2+(nd)^2]^{\frac12}-nd\}) \cos(nx)$

$$= d^{-\frac12} \sum_{m_1}^{m_2} |2\pi m+x|^{-\frac12}[a-|(2\pi m+x)/d|]^{-\frac12} J_{2\nu}\{b|(2\pi m+x)/d|^{\frac12}[a-|(2\pi m+x)/d|^{\frac12}]\}$$

$\operatorname{Re}\nu>-\frac12$; $\operatorname{Re}\nu>\frac12$ if $x=0$ or $x=\pm ad$

4.85 $\displaystyle\sum_{0}^{\infty} \epsilon_n J_{\nu-(and/\pi)}(y) J_{\nu+(and/\pi)}(y) \cos(nx) = \pi(ad)^{-1} \sum_{m_1}^{m_2} J_{2\nu}\{2y\cos[(\pi/2ad)(2\pi m+x)]\}$

$\operatorname{Re}\nu>-\frac12$; $\operatorname{Re}\nu>0$ if $x=\pm ad$

4.86 $\displaystyle\sum_{0}^{\infty} \epsilon_n \cos(\tfrac12 and) J_{\nu-(and/2\pi)}(y) J_{\nu+(and/2\pi)}(y) \cos(nx)$

$$= \pi(ad)^{-1} \sum_{m_1}^{m_2} J_{2\nu}\{2y\sin[(\pi/ad)(2\pi m+x)]\}$$

$\operatorname{Re}\nu>-\frac12$; $\operatorname{Re}\nu>0$ if $x=0$ or $x=\pm ad$

4.87 $\displaystyle\sum_{0}^{\infty} \epsilon_n I_{\nu-(and/\pi)}(y) I_{\nu+(and/\pi)}(y) \cos(nx) = \pi(ad)^{-1} \sum_{m_1}^{m_2} I_{2\nu}\{2y\cos[(\pi/2ad)(2\pi m+x)]\}$

$\operatorname{Re}\nu>-\frac12$; $\operatorname{Re}\nu>0$ if $x=\pm ad$

4.88 $\displaystyle\sum_{0}^{\infty} \epsilon_n \cos(\tfrac12 and) I_{\nu-(and/\pi)}(y) I_{\nu+(and/\pi)}(y) \cos(nx)$

$$= \pi(ad)^{-1} \sum_{m_1}^{m_2} I_{2\nu}\{2y\sin[(\pi/ad)(2\pi m+x)]\}, \quad \operatorname{Re}\nu>-\tfrac12;\ \operatorname{Re}\nu>0 \text{ if } x=0 \text{ or } \pm ad$$

4.89 $\displaystyle\sum_{0}^{\infty} \epsilon_n (nd)^{-\nu-1} \mathbf{H}_\nu(and) \cos(nx)$

$$= 2(2\pi/a)^{\frac12} d^{-1} \sum_{m_1}^{m_2} \{a^2-[(2\pi m+x)/d]^2\}^{\frac12\nu+\frac14} P_{\nu-\frac12}^{-\nu-\frac12}[|(2\pi m+x)/ad|]$$

$\operatorname{Re}\nu>-\frac32$; $\operatorname{Re}\nu>-\frac12$ if $x=\pm ad$

4.90 $\displaystyle\sum_{0}^{\infty} \epsilon_n (nd)^{-\mu-1} s_{\mu,\nu}(and) \cos(nx)$

$$= 2^{\frac32+\mu}(\pi/a)^{\frac12} d^{-1}[(\mu+1)^2-\nu^2]^{-1} \Gamma[\tfrac12(3+\mu+\nu)]\Gamma[\tfrac12(3+\mu-\nu)]$$

$$\cdot \sum_{m_1}^{m_2} \{a^2-[(2\pi m+x)/d]^2\}^{\frac12\mu+\frac14} P_{\nu-\frac12}^{-\mu-\frac12}[|(2\pi m+x)/ad|]$$

$\operatorname{Re}\mu>-\frac32$; $\operatorname{Re}\mu>-\frac12$ if $x=\pm ad$

4.91 $$\sum_{0}^{\infty} \epsilon_n P_\nu^{and/\pi}(y) P_\nu^{-and/\pi}(y) \cos(nx)$$
$$= (\pi/ad) \sum_{m_1}^{m_2} P_\nu\{1-2(1-y^2)\cos^2[(\pi/2ad)(2\pi m+x)]\}, \quad 0<y<1, \quad x\neq\pm ad$$

4.92 $$\sum_{0}^{\infty} \epsilon_n P_\nu^{and/\pi}(z) P_\nu^{-and/\pi}(z) \cos(nx)$$
$$= (\pi/ad) \sum_{m_1}^{m_2} P_\nu\{1+2(z^2-1)\cos^2[(\pi/2ad)(2\pi m+x)]\}, \quad z>1, \quad x\neq\pm ad$$

4.93 $$\sum_{0}^{\infty} \epsilon_n \cos(yn) Q_{-\frac{1}{2}+(and/\pi)}(z) Q_{-\frac{1}{2}-(and/\pi)}(z) \cos(nx)$$
$$= (\pi^2/ad) \sum_{m_1}^{m_2} \{y^2-\sin^2[(\pi/2ad)(2\pi m+x)]\}^{-\frac{1}{2}}$$
$$\cdot K(\cos[(\pi/2ad)(2\pi m+x)]\{y^2-\sin^2[(2\pi^2 m+\pi x)/2ad]\}^{-\frac{1}{2}}), \quad z>1$$

4.94 $$\sum_{0}^{\infty} \epsilon_n [P_{-\frac{1}{2}+\frac{1}{2}ind}(\cosh a)]^2 \cos(nx) = (4/\pi d \sinh a) \sum_{m_1}^{m_2} K'\left(\left|\frac{\sinh[(2\pi m+x)/d]}{\sinh a}\right|\right), \quad a>0$$

4.95 $$\sum_{0}^{\infty} \epsilon_n P^{\mu}_{-\frac{1}{2}+\frac{1}{2}ind}(\cosh a) P^{-\mu}_{-\frac{1}{2}+\frac{1}{2}ind}(\cosh a) \cos(nx)$$
$$= \frac{2}{d \sinh a} \sum_{m_1}^{m_2} P_{\mu-\frac{1}{2}}\left[2\left(\frac{\sinh(2\pi m+x/d)}{\sinh a}\right)^2 - 1\right], \quad a>0, \quad x\neq\pm ad$$

4.96 $$\sum_{0}^{\infty} \epsilon_n D_{\nu+(2and/\pi)}(y) D_{\nu-(2and/\pi)}(y) \cos(nx)$$
$$= 2^{-\nu}(\tfrac{1}{2}\pi)^{\frac{3}{2}} \exp(\tfrac{1}{4}y^2)(ad)^{-1} \sum_{m_1}^{m_2} \{\cos[(\pi/2ad)(2\pi m+x)]\}^{-\nu-1}$$
$$\cdot \exp\{-\tfrac{1}{4}y^2 \sec[(\pi/2ad)(2\pi m+x)]\} D_{2\nu+1}(y\{1+\sec[(\pi/2ad)(2\pi m+x)]\}^{\frac{1}{2}})$$

In 4.75–4.96, $m_1 = -[(ad+x)/2\pi]$; $m_2 = [(ad-x)/2\pi]$. If $(ad\pm x)/2\pi$ is an integer or zero, one half of the corresponding term in the right side sum has to be taken.

4.97 $$\sum_{0}^{\infty} (-1)^n \epsilon_n \Gamma(\tfrac{1}{2}+2n) D_{-2n-\frac{1}{2}}[(\tfrac{1}{2}ia^2)^{\frac{1}{2}}] D_{-2n-\frac{1}{2}}[(-\tfrac{1}{2}ia^2)^{\frac{1}{2}}] \cos(2nx)$$
$$= \pi(\cos x)^{-\frac{1}{2}}\{\cos^2(a^2/4\cos x)[\tfrac{1}{2}-S(a^2/4\cos x)] - \sin(a^2/4\cos x)[\tfrac{1}{2}-C(a^2/4\cos x)]\}$$

4.98 $\sum_0^\infty (-1)^n \Gamma(\frac{3}{2}+2n) D_{-2n-\frac{3}{2}}[(\frac{1}{2}ia^2)^{\frac{1}{2}}] D_{-2n-\frac{3}{2}}[(-\frac{1}{2}ia^2)^{\frac{1}{2}}] \cos[(2n+1)x]$

$$= \tfrac{1}{2}\pi(\cos x)^{-\frac{1}{2}}\{\cos(a^2/4\cos x)[\tfrac{1}{2}-C(a^2/4\cos x)]+\sin(a^2/4\cos x)[\tfrac{1}{2}-S(a^2/4\cos x)]\}$$

4.99 $\sum_0^\infty (-1)^n \epsilon_n J_{2n}\{z[(a^2+1)^{\frac{1}{2}}-a]^{\frac{1}{2}}\} K_{2n}\{z[(a^2+1)^{\frac{1}{2}}+a]^{\frac{1}{2}}\} \cos(2nx)$

$$= \tfrac{1}{2}K_0[z(a+i\cos x)^{\frac{1}{2}}]+\tfrac{1}{2}K_0[z(a-i\cos x)^{\frac{1}{2}}]$$

4.100 $\sum_0^\infty (-1)^n J_{2n+1}\{z[(a^2+1)^{\frac{1}{2}}-a]^{\frac{1}{2}}\} K_{2n+1}\{z[(a^2+1)^{\frac{1}{2}}+a]^{\frac{1}{2}}\} \cos[(2n+1)x]$

$$= \tfrac{1}{4}iK_0[z(a+i\cos x)^{\frac{1}{2}}]-\tfrac{1}{4}iK_0[z(a-i\cos x)^{\frac{1}{2}}]$$

4.101 $\sum_0^\infty (-1)^n \epsilon_n [P_{n-\frac{1}{2}}(y)]^2 \cos(2nx) = (2/\pi)(1-y^2)^{-\frac{1}{2}}K'[\cos x(1-y^2)^{-\frac{1}{2}}], \quad \cos x<(1-y^2)^{\frac{1}{2}}$

$$= 0, \qquad \cos x>(1-y^2)^{\frac{1}{2}}$$

$$0<y<1$$

4.102 $\sum_0^\infty (-1)^n [P_n(y)]^2 \cos[(2n+1)x] = \pi^{-1}(1-y^2)^{-\frac{1}{2}}K[(1-y^2)^{-\frac{1}{2}}\cos x], \quad \cos x<(1-y^2)^{\frac{1}{2}}$

$$= (\pi\cos x)^{-1}K[(1-y^2)^{-\frac{1}{2}}\sec x], \qquad \cos x>(1-y^2)^{\frac{1}{2}}$$

$$0<y<1$$

4.103 $\sum_0^\infty (-1)^n \epsilon_n [\Gamma(n-\mu+\frac{1}{2})/\Gamma(n+\mu+\frac{1}{2})][P^{\mu}_{n-\frac{1}{2}}(y)]^2 \cos(2nx)$

$$= (2/\pi)(1-y^2)^{-\frac{1}{2}}Q_{-\mu-\frac{1}{2}}\{1-[2\cos^2 x/(1-y^2)]\}, \qquad 0<\cos x<(1-y^2)^{\frac{1}{2}}$$

$$= (2/\pi)(1-y^2)^{-\frac{1}{2}}\sin(\pi\mu)Q_{-\mu-\frac{1}{2}}\{[2\cos^2 x/(1-y^2)]-1\}, \quad (1-y^2)^{\frac{1}{2}}<\cos x<1$$

$$0<y<1$$

4.104 $\sum_0^\infty (-1)^n \epsilon_n P^{\mu}_{n-\frac{1}{2}}(y) P^{-\mu}_{n-\frac{1}{2}}(y) \cos(2nx)$

$$= (1-y^2)^{-\frac{1}{2}}P_{\mu-\frac{1}{2}}\{[2\cos^2 x/(1-y^2)]-1\}, \quad 0<\cos x<(1-y^2)^{\frac{1}{2}}$$

4.105 $\sum_0^\infty (-1)^n \epsilon_n [\Gamma(\frac{1}{2}-\mu+n)/\Gamma(\frac{1}{2}+\mu+n)] P^{\mu}_{n-\frac{1}{2}}(z) Q^{\mu}_{n-\frac{1}{2}}(z) \cos(2nx)$

$$= (z^2-1)^{-\frac{1}{2}}e^{i\pi\mu}Q_{-\mu-\frac{1}{2}}\{1+[2\cos^2 x/(z^2-1)]\}, \quad z>1$$

4.106 $\sum_0^\infty (-1)^n \epsilon_n P_{n-\frac{1}{2}}(z) Q_{n-\frac{1}{2}}(z) \cos(2nx) = (z^2-\sin^2 x)^{-\frac{1}{2}}K[(z^2-1)^{\frac{1}{2}}(z^2-\sin^2 x)^{-\frac{1}{2}}], \quad z>1$

4.107 $\sum_{0}^{\infty} (-1)^n \epsilon_n [Q_{n-\frac{1}{2}}(z)]^2 \cos(2nx) = \pi (z^2 - \sin^2 x)^{-\frac{1}{2}} K[\cos x (z^2 - \sin^2 x)^{-\frac{1}{2}}], \quad z > 1$

4.108 $\sum (-1)^n P_n(z) Q_n(z) \cos[(2n+1)x] = \frac{1}{2}(z^2 - \sin^2 x)^{-\frac{1}{2}} K[\cos x (z^2 - \sin^2 x)^{-\frac{1}{2}}], \quad z > 1$

4.109 $\sum_{0}^{\infty} \epsilon_n P_\nu^{\,n}(y) P_\nu^{-n}(y) \cos[n\pi(x/a)] = P_\nu\{1 - 2(1-y^2)\cos^2[\pi(x/a)]\}, \quad 0 < y < 1$

4.110 $\sum_{0}^{\infty} \epsilon_n P_\nu^{\,n}(z) P_\nu^{-n}(z) \cos[n\pi(x/a)] = P_\nu\{1 + 2(z^2-1)\cos^2[\pi(x/a)]\}, \quad z > 1$

4.111 $\sum_{0}^{\infty} (-1)^n \epsilon_n Q_{-\frac{1}{2}+n}(z) Q_{-\frac{1}{2}-n}(z) \cos[n\pi(x/a)]$

$= \pi[z^2 - \sin^2(\pi x/2a)]^{-\frac{1}{2}} K\{\cos(\pi x/2a)[z^2 - \sin^2(\pi x/2a)]^{-\frac{1}{2}}\}, \quad z > 1$

4.112 $\sum_{0}^{\infty} (-1)^n [\Gamma(n-\mu+1)/\Gamma(n+\mu+1)][P_n^{\,\mu}(y)]^2 \cos[(2n+1)x]$

$= \frac{1}{2}(1-y^2)^{-\frac{1}{2}} P_{-\mu-\frac{1}{2}}\{1 - [2\cos^2 x/(1-y^2)]\}, \qquad \cos x < (1-y^2)^{\frac{1}{2}}$

$= \pi^{-1}(1-y^2)^{-\frac{1}{2}} Q_{-\mu-\frac{1}{2}}\{[2\cos^2 x/(1-y^2)] - 1\} \cos(\pi\mu), \quad \cos x > (1-y^2)^{\frac{1}{2}}$

$\operatorname{Re}\mu < 1$

4.113 $\sum_{0}^{\infty} (-1)^n \epsilon_n [\Gamma(\frac{1}{2}+n-\mu)/\Gamma(\frac{1}{2}+n+\mu)][Q_{n-\frac{1}{2}}^{\mu}(z)]^2 \cos(2nx)$

$= \frac{1}{2}\pi^2 (z^2-1)^{-\frac{1}{2}} \sec(\pi\mu) P_{-\mu-\frac{1}{2}}\{1 + [2\cos^2 x/(z^2-1)]\} \exp(i2\pi\mu), \quad z > 1$

4.114 $\sum_{0}^{\infty} (-1)^n \epsilon_n P_\nu^{\,n}(\cos\vartheta_1) P_\nu^{-n}(\cos\vartheta_2) \cos(nx)$

$= \sum_{0}^{\infty} \epsilon_n [\Gamma(\nu-n+1)/\Gamma(\nu+n+1)] P_\nu^{\,n}(\cos\vartheta_1) P_\nu^{\,n}(\cos\vartheta_2) \cos(nx)$

$= P_\nu(\cos\vartheta_1 \cos\vartheta_2 + \sin\vartheta_1 \sin\vartheta_2 \cos x), \quad 0 \le \vartheta_{\substack{1\\2}} < \pi, \quad \vartheta_1 + \vartheta_2 < \pi$

4.115 $\sum_{0}^{\infty} (-1)^n \epsilon_n P_\nu^{-n}(\cos\vartheta_1) Q_\nu^{\,n}(\cos\vartheta_2) \cos(nx)$

$= \sum_{0}^{\infty} \epsilon_n [\Gamma(\nu-n+1)/\Gamma(\nu+n+1)] P_\nu^{\,n}(\cos\vartheta_1) Q_\nu^{\,n}(\cos\vartheta_2) \cos(nx)$

$= Q_\nu(\cos\vartheta_1 \cos\vartheta_2 + \sin\vartheta_1 \sin\vartheta_2 \cos x), \quad 0 < \vartheta_2 < \frac{1}{2}\pi, \quad 0 < \vartheta_1 < \pi, \quad 0 < \vartheta_1 + \vartheta_2 < \pi$

4.116 $$\sum_0^\infty (-1)^n \epsilon_n P_\nu^n(z_1) P_\nu^n(z_2) [\Gamma(\nu-n+1)/\Gamma(\nu+n+1)] \cos(nx)$$
$$= \sum_0^\infty (-1)^n \epsilon_n P_\nu^n(z_1) P_\nu^{-n}(z_2) \cos(nx)$$
$$P_\nu[z_1 z_2 - (z_1^2-1)^{\frac{1}{2}}(z_2^2-1)^{\frac{1}{2}} \cos x], \quad \mathrm{Re}\, z_{1,2} > 0, \quad |\arg(z_{1,2}-1)| < \pi$$

4.117 $$\sum_0^\infty (-1)^n \epsilon_n Q_\nu^n(z_1) P_\nu^{-n}(z_2) \cos(nx)$$
$$= \sum_0^\infty (-1)^n \epsilon_n [\Gamma(\nu-n+1)/\Gamma(\nu+n+1)] Q_\nu^n(z_1) P_\nu^n(z_2) \cos(nx)$$
$$= Q_\nu[z_1 z_2 - (z_1^2-1)^{\frac{1}{2}}(z_2^2-1)^{\frac{1}{2}} \cos x], \quad z_{1,2} \text{ real}, \quad 1 < z_2 < z_1, \quad \nu \neq -1, -2, -3, \ldots$$

4.118 $$\sum_0^\infty P_\nu^{n+\frac{1}{2}}(z) P_\nu^{-n-\frac{1}{2}}(z) \cos[(2n+1)x] = \tfrac{1}{2} P_\nu[z^2 + (z^2-1)\cos(2x)], \quad z > 1$$

4.119 $$\sum_0^\infty P_\nu^{n+\frac{1}{2}}(y) P_\nu^{-n-\frac{1}{2}}(y) \cos[(2n+1)x] = \tfrac{1}{2} P_\nu[y^2 - (1-y^2)\cos(2x)], \quad 0 < y < 1$$

4.120 $$\sum_0^\infty \epsilon_n [\Gamma(1+\nu-nm)/\Gamma(1+\nu+nm)][P_\nu^{nm}(y)]^2 \cos(nx)$$
$$= m^{-1} \sum_{r_1}^{r_2} P_\nu\{y^2 + (1-y^2)\cos[(2\pi r+x)/m]\}, \quad 0 < y < 1$$

4.121 $$\sum_0^\infty (-1)^{nm} \epsilon_n [\Gamma(1+\nu-nm)/\Gamma(1+\nu+nm)][P_\nu^{nm}(z)]^2 \cos(nx)$$
$$= m^{-1} \sum_{r_1}^{r_2} P_\nu\{z^2 - (z^2-1)\cos[(2\pi r+x)/m]\}, \quad z > 1$$

4.122 $$\sum_0^\infty (-1)^{nm} \epsilon_n [\Gamma(1+\nu-nm)/\Gamma(1+\nu+nm)] P_\nu^{nm}(z) Q_\nu^{nm}(z) \cos(nx)$$
$$= m^{-1} \sum_{r_1}^{r_2} Q_\nu\{z^2 - (z^2-1)\cos[(2\pi r+x)/m]\}, \quad z > 1$$

In the last three formulas $m = 1, 2, 3, \ldots$, $r_1 = -[(m\pi + x)/2\pi]$, $r_2 = [(m\pi - x)/2\pi]$. If $(m\pi \pm x)/2\pi$ is an integer or zero, one half of the corresponding term in the sum at the right side has to be taken.

V

EXPONENTIAL FOURIER AND FOURIER–BESSEL SERIES

In formulas 5.1–5.33 the properties m_1, m_2, and $H(m)$ on the right-hand sides are

$$m_1 = -[(ad+x)/2\pi], \qquad m_2 = [(ad-x)/2\pi]; \qquad H(m) = \exp[-iy(2\pi m+x)/d]$$

If $(ad \pm x)/2\pi$ is an integer or zero, one half of the corresponding term in the sum has to be taken.

5.1 $$\sum_{-\infty}^{\infty} \exp(inx)[b^2+(y+nd)^2]^{-\frac{1}{2}} \sin\{a[b^2+(y+nd)^2]^{\frac{1}{2}}\}$$
$$= (\pi/d) \sum_{m_1}^{m_2} H(m) J_0(b\{a^2-[(2\pi m+x)/d]^2\}^{\frac{1}{2}})$$

5.2 $$\sum_{-\infty}^{\infty} \exp(inx) \cos[\tfrac{1}{2}a(y+nd)]\{\Gamma[\nu+(ay/2\pi)+(and/2\pi)]\Gamma[\nu-(ay/2\pi)-(and/2\pi)]\}^{-1}$$
$$= \pi 2^{2\nu-2}[ad\Gamma(2\nu-1)]^{-1} \sum_{m_1}^{m_2} H(m) \,|\sin[(\pi/ad)(2\pi m+x)]|^{2\nu-2}$$

$\mathrm{Re}\nu > \frac{1}{2}$; $\mathrm{Re}\nu > 1$ if $x = \pm ad$ or $x = 0$

5.3 $$\sum_{-\infty}^{\infty} \exp(inx)\{\Gamma[\nu+(a/\pi)y+(a/\pi)nd]\Gamma[\nu-(a/\pi)y-(a/\pi)nd]\}^{-1}$$
$$=2^{2\nu-2}\pi[ad\Gamma(2\nu-1)]^{-1}\sum_{m_1}^{m_2} H(m)\{\cos[(\pi/2ad)(2\pi m+x)]\}^{2\nu-2}$$
$\mathrm{Re}\nu>\frac{1}{2}$; $\mathrm{Re}\nu>1$ if $x=\pm ad$

5.4 $$\sum_{-\infty}^{\infty} \exp(inx)P^{\mu}_{-\frac{1}{2}+y+nd}(\cos a)$$
$$=[(2\pi)^{\frac{1}{2}}/d\Gamma(\tfrac{1}{2}-\mu)](\sin a)^{\mu}\sum_{m_1}^{m_2} H(m)\{\cos[(2\pi m+x)/d]-\cos a\}^{-\mu-\frac{1}{2}}$$
$0<a<\pi$, $\mathrm{Re}\mu<\frac{1}{2}$; $\mathrm{Re}\mu<-\frac{1}{2}$ if $x=\pm ad$

5.5 $$\sum_{-\infty}^{\infty} \exp(inx)P_{-\frac{1}{2}+iy+ind}(\cosh a)$$
$$=[(2\pi)^{\frac{1}{2}}/d\Gamma(\tfrac{1}{2}-\mu)](\sinh a)^{\mu}\sum_{m_1}^{m_2} H(m)\{\cosh a-\cosh[(2\pi m+x)/d]\}^{-\mu-\frac{1}{2}}$$
$\mathrm{Re}\mu<\frac{1}{2}$; $\mathrm{Re}\mu<-\frac{1}{2}$ if $x=\pm ad$

5.6 $$\sum_{-\infty}^{\infty} \exp(inx)P_{\nu}^{a(y+nd)/\pi}(b)P_{\nu}^{-a(y+nd)/\pi}(b)$$
$$=(\pi/ad)\sum_{m_1}^{m_2} H(m)P_{\nu}\{1-2(1-b^2)\cos^2[(\pi/2ad)(2\pi m+x)]\},\quad 0<b<1,\quad x\neq\pm ad$$

5.7 $$\sum_{-\infty}^{\infty} \exp(inx)P_{\nu}^{a(y+nd)/\pi}(b)P_{\nu}^{-a(y+nd)/\pi}(b)$$
$$=(\pi/ad)\sum_{m_1}^{m_2} H(m)P_{\nu}\{1+2(b^2-1)\cos^2[(\pi/2ad)(2\pi m+x)]\},\quad b>1,\quad x\neq\pm ad$$

5.8 $$\sum_{-\infty}^{\infty} \exp(inx)\cos(an)Q_{-\frac{1}{2}+a(y+nd)/\pi}(b)Q_{-\frac{1}{2}-a(y+nd)/\pi}(b)$$
$$=(\pi^2/ad)\sum_{m_1}^{m_2} H(m)\{b^2-\sin^2[(\pi/2ad)(2\pi m+x)]\}^{-\frac{1}{2}}$$
$$\cdot K(\cos[(\pi/2ad)(2\pi m+x)]\{b^2-\sin^2[(2\pi^2 m+\pi x)/2ad]\}^{-\frac{1}{2}}),\quad b>1$$

5.9 $$\sum_{-\infty}^{\infty} \exp(inx)P^{\mu}_{-\frac{1}{2}+\frac{1}{2}iy+\frac{1}{2}ind}(\cosh a)P^{-\mu}_{-\frac{1}{2}+\frac{1}{2}iy+\frac{1}{2}ind}(\cosh a)$$
$$=(2/d\sinh a)\sum_{m_1}^{m_2} H(m)P_{\mu-\frac{1}{2}}\left(2\frac{\sinh^2[(2\pi m+x)/d]}{\sinh^2 a}-1\right),\quad x\neq 0,\quad x\neq\pm ad$$

5.10 $\sum_{-\infty}^{\infty} \exp(inx)(y+nd)^{-\nu} J_\nu[a(y+nd)]$

$$= [\pi^{\frac{1}{2}} 2^{1-\nu} a^{-\nu}/d\Gamma(\tfrac{1}{2}+\nu)] \sum_{m_1}^{m_2} H(m)\{a^2-[(2\pi m+x)/d]^2\}^{\nu-\frac{1}{2}}$$

$\operatorname{Re}\nu > -\frac{1}{2}$; $\operatorname{Re}\nu > \frac{1}{2}$ if $x = \pm ad$

5.11 $\sum_{-\infty}^{\infty} \exp(inx)(y+nd)^{-\nu} J_{\nu+2l}[a(y+nd)]$

$$= [(-1)^l (\tfrac{1}{2}a)^{-\nu}\Gamma(\nu)(2l)!/d\Gamma(2l+2\nu)] \sum_{m_1}^{m_2} H(m)\{a^2-[(2\pi m+x)/d]^2\}^{\nu-\frac{1}{2}}$$

$\cdot C_{2l}^{\nu}[(2\pi m+x)/ad]$, $l = 0, 1, 2, \ldots$; $\operatorname{Re}\nu > -\frac{1}{2}$; $\operatorname{Re}\nu > \frac{1}{2}$ if $x = \pm ad$

5.12 $\sum_{-\infty}^{\infty} \exp(inx) \cos[\tfrac{1}{2}a(y+nd)] J_\nu[\tfrac{1}{2}a(y+nd)](y+nd)^{-\nu}$

$$= [\pi^{\frac{1}{2}} a^{-\nu}/d\Gamma(\tfrac{1}{2}+\nu)] \sum_{m_1}^{m_2} H(m) \,|(2\pi m+x)/d|^{\nu-\frac{1}{2}} [a-|(2\pi m+x)/d|]^{\nu-\frac{1}{2}}$$

$\operatorname{Re}\nu > -\frac{1}{2}$; $\operatorname{Re}\nu > \frac{1}{2}$ if $x = 0$ and $x = \pm ad$

5.13 $\sum_{-\infty}^{\infty} \exp(inx) J_\nu[\tfrac{1}{2}a(y+nd)] J_{-\nu}[\tfrac{1}{2}a(y+nd)]$

$$= 2(ad)^{-1} \sum_{m_1}^{m_2} H(m) P_{\nu-\frac{1}{2}}[2(ad)^{-1}(2\pi m+x)^2 - 1]$$

5.14 $\sum_{-\infty}^{\infty} \exp(inx)(y+nd)^{\frac{1}{2}} J_\nu[\tfrac{1}{2}a(y+nd)] J_{-\nu-\frac{1}{2}}[\tfrac{1}{2}a(y+nd)]$

$$= 2(\tfrac{1}{2}\pi d)^{-\frac{1}{2}} \sum_{m_1}^{m_2} H(m)\,|2\pi m+x|^{-\frac{1}{2}} \{a^2-[(2\pi m+x)/d]^2\}^{-\frac{1}{2}}$$

$\cdot \cos\{(2\nu+\tfrac{1}{2}) \arccos[(2\pi m+x)/ad]\}$, $x \neq 0$, $\pm ad$

5.15 $\sum_{-\infty}^{\infty} \exp(inx) \cos[\tfrac{1}{2}a(y+nd)] J_0\{\tfrac{1}{2}a[(y+nd)^2+b^2]^{\frac{1}{2}}\}$

$$= d^{-\frac{1}{2}} \sum_{m_1}^{m_2} H(m)\,|2\pi m+x|^{-\frac{1}{2}} [a-|(2\pi m+x)/d|]^{-\frac{1}{2}}$$

$\cdot \cos\{b[(2\pi m+x)/d]^{\frac{1}{2}}[a-|(2\pi m+x)/d|^{\frac{1}{2}}]^{\frac{1}{2}}\}$, $x \neq 0$, $\pm ad$

5.16 $\sum_{-\infty}^{\infty} \exp(inx) J_0\{a[(y+nd)^2+b^2]^{\frac{1}{2}}\}$

$$= (2/d) \sum_{m_1}^{m_2} H(m)\{a^2-[(2\pi m+x)/d]^2\}^{-\frac{1}{2}} \cos(b\{a^2-[(2\pi m+x)/d]^2\}^{\frac{1}{2}}), \quad x \neq \pm ad$$

5.17 $\sum_{-\infty}^{\infty} \exp(inx) J_0\{a[(y+nd)^2-b^2]^{\frac{1}{2}}\}$

$= (2/d) \sum_{m_1}^{m_2} H(m) \{a^2-[(2\pi m+x)/d]^2\}^{-\frac{1}{2}} \cosh(b\{a^2-[(2\pi m+x)/d]^2\}^{\frac{1}{2}}), \quad x \neq \pm ad$

5.18 $\sum_{-\infty}^{\infty} \exp(inx) [b^2+(y+nd)^2]^{-\frac{1}{2}\nu} J_\nu\{a[b^2+(y+nd)^2]^{\frac{1}{2}}\}$

$= [(2\pi b)^{\frac{1}{2}}/d](ab)^{-\nu} \sum_{m_1}^{m_2} H(m) \{a^2-[(2\pi m+x)/d]^2\}^{\frac{1}{2}\nu-\frac{1}{4}} J_{\nu-\frac{1}{2}}(b\{a^2-[(2\pi m+x)/d]^2\}^{\frac{1}{2}})$

$\mathrm{Re}\,\nu > -\frac{1}{2}$; $\mathrm{Re}\,\nu > \frac{1}{2}$ if $x = \pm ad$

5.19 $\sum_{-\infty}^{\infty} \exp(inx) T_{2l}\{(y+nd)/[b^2+(y+nd)^2]^{\frac{1}{2}}\} J_{2l}\{a[b^2+(y+nd)^2]^{\frac{1}{2}}\}$

$= (-1)^l (2/d) \sum_{m_1}^{m_2} H(m) \{a^2-[(2\pi m+x)/d]^2\}^{-\frac{1}{2}}$

$\cdot \cos(b\{a^2-[(2\pi m+x)/d]^2\}^{\frac{1}{2}}) T_{2l}[(2\pi m+x)/ad], \quad l=0, 1, 2, \ldots, \quad x \neq ad$

5.20 $\sum_{-\infty}^{\infty} \exp(inx) T_{2l}\{b[b^2+(y+nd)^2]^{-\frac{1}{2}}\} J_{2l}\{a[b^2+(y+nd)^2]^{\frac{1}{2}}\}$

$= (-1)^l (2/d) \sum_{m_1}^{m_2} H(m) \{a^2-[(2\pi m+x)/d]^2\}^{-\frac{1}{2}} \cos(b\{a^2-[(2\pi m+x)/d]^2\}^{\frac{1}{2}})$

$\cdot T_{2l}(\{1-[(2\pi m+x)/ad]^2\}^{\frac{1}{2}}), \quad l=0, 1, 2, \ldots, \quad x \neq ad$

5.21 $\sum_{-\infty}^{\infty} \exp(inx) T_{2l+1}\{b[b^2+(y+nd)^2]^{-\frac{1}{2}}\} J_{2l+1}\{a[b^2+(y+nd)^2]^{\frac{1}{2}}\}$

$= (-1)^l (2/d) \sum_{m_1}^{m_2} H(m) \{a^2-[(2\pi+x)/d]^2\}^{-\frac{1}{2}} \sin(b\{a^2-[(2\pi m+x)/d]^2\}^{\frac{1}{2}})$

$\cdot T_{2l+1}(\{1-[(2\pi m+x)/ad]^2\}^{\frac{1}{2}}), \quad l=0, 1, 2, \ldots$

5.22 $\sum_{-\infty}^{\infty} \exp(inx) [b^2+(y+nd)^2]^{-\frac{1}{2}} U_{2l}\{(y+nd)/[b^2+(y+nd)^2]^{\frac{1}{2}}\} J_{2l+1}\{a[b^2+(y+nd)^2]^{\frac{1}{2}}\}$

$= (-1)^l 2(abd)^{-1} \sum_{m_1}^{m_2} H(m) \sin(b\{a^2-[(2\pi m+x)/d]^2\}^{\frac{1}{2}})$

$\cdot U_{2l}[(2\pi m+x)/ad], \quad l=0, 1, 2, \ldots$

5.23 $\sum_{-\infty}^{\infty} \exp(inx) J_\nu(\tfrac{1}{2}a\{[b^2+(y+nd)^2]^{\frac{1}{2}}+y+nd\}) J_\nu(\tfrac{1}{2}a\{[b^2+(y+nd)^2]^{\frac{1}{2}}-y-nd\})$

$= (2/d) \sum_{m_1}^{m_2} H(m)\{a^2-[(2\pi m+x)/d]^2\}^{-\frac{1}{2}} J_{2\nu}(b\{a^2-[(2\pi m+x)/d]^2\}^{\frac{1}{2}})$

$\mathrm{Re}\,\nu > -\tfrac{1}{2}$; $\mathrm{Re}\,\nu > \tfrac{1}{2}$ if $x = \pm ad$

5.24 $\sum_{-\infty}^{\infty} \exp(inx) \cos[\tfrac{1}{2}a(y+nd)] J_\nu(\tfrac{1}{4}a\{[b^2+(y+nd)^2]^{\frac{1}{2}}+y+nd\})$

$\cdot J_\nu(\tfrac{1}{4}a\{[b^2+(y+nd)^2]^{\frac{1}{2}}-y-nd\})$

$= d^{-\frac{1}{2}} \sum_{m_1}^{m_2} H(m) \,|2\pi m+x|^{-\frac{1}{2}} [a-|(2\pi m+x)/d|]^{-\frac{1}{2}}$

$\cdot J_{2\nu}\{b\,|(2\pi m+x)/d|^{\frac{1}{2}} [a-|(2\pi m+x)/d|]^{\frac{1}{2}}\}$

$\mathrm{Re}\,\nu > -\tfrac{1}{2}$; $\mathrm{Re}\,\nu > \tfrac{1}{2}$ if $x = 0$ or $\pm ad$

5.25 $\sum_{-\infty}^{\infty} \exp(inx) J_{\nu-[a(y+nd)/\pi]}(b) J_{\nu+[a(y+nd)/\pi]}(b)$

$= (\pi/ad) \sum_{m_1}^{m_2} H(m) J_{2\nu}\{2b \cos[(\pi/2ad)(2\pi m+x)]\}$, $\mathrm{Re}\,\nu > -\tfrac{1}{2}$; $\mathrm{Re}\,\nu > 0$ if $x = \pm ad$

5.26 $\sum_{-\infty}^{\infty} \exp(inx) \cos[\tfrac{1}{2}a(y+nd)] J_{\nu-[a(y+nd)/2\pi]}(b) J_{\nu+[a(y+nd)/\pi]}(b)$

$= (\pi/ad) \sum_{m_1}^{m_2} H(m) J_{2\nu}\{2b \sin[(\pi/ad)(2\pi m+x)]\}$

$\mathrm{Re}\,\nu > -\tfrac{1}{2}$; $\mathrm{Re}\,\nu > 0$ if $x = 0$ or $\pm ad$

5.27 $\sum_{-\infty}^{\infty} \exp(inx) I_{\nu-[a(y+nd)/\pi]}(b) I_{\nu+[a(y+nd)/\pi]}(b)$

$= (\pi/ad) \sum_{m_1}^{m_2} H(m) I_{2\nu}\{2b \cos[(\pi/2ad)(2\pi m+x)]\}$, $\mathrm{Re}\,\nu > -\tfrac{1}{2}$; $\mathrm{Re}\,\nu > 0$ if $x = \pm ad$

5.28 $\sum_{-\infty}^{\infty} \exp(inx) \cos[\tfrac{1}{2}a(y+nd)] I_{\nu-[a(y+nd)/2\pi]}(b) I_{\nu+[a(y+nd)/2\pi]}(b)$

$= (\pi/ad) \sum_{m_1}^{m_2} H(m) I_{2\nu}\{2b \sin[(\pi/ad)(2\pi m+x)]\}$

$\mathrm{Re}\,\nu > -\tfrac{1}{2}$; $\mathrm{Re}\,\nu > 0$ if $x = 0$ or $\pm ad$

5.29 $\sum_{-\infty}^{\infty} \exp(inx)\{\mathbf{J}_\nu[a(y+nd)]+\mathbf{J}_{-\nu}[a(y+nd)]\}$

$$=(4/d)\cos(\tfrac{1}{2}\pi\nu)\sum_{m_1}^{m_2} H(m)\{a^2-[(2\pi m+x)/d]\}^{-\frac{1}{2}}\cos\{\nu \arccos[(2\pi m+x)/ad]\}$$

$x\neq\pm ad$

5.30 $\sum_{-\infty}^{\infty} \exp(inx)[\mathbf{J}_{a(y+nd)/\pi}(b)+\mathbf{J}_{-a(y+nd)/\pi}(b)]$

$$=(2\pi/ad)\sum_{m_1}^{m_2} H(m)\cos\{b\sin[(\pi/ad)(2\pi m+x)]\}$$

5.31 $\sum_{-\infty}^{\infty} \exp(inx)[E_{a(y+nd)/\pi}(b)+E_{-a(y+nd)/\pi}(b)]$

$$=(-2\pi/ad)\sum_{m_1}^{m_2} H(m)\sin\{b\sin[(\pi/ad)(2\pi m+x)]\}$$

$\sum_{-\infty}^{\infty} \exp(inx)[\cos(ay+and)]^{-1}[\mathbf{J}_{2a(y+nd)/\pi}(b)+\mathbf{J}_{-2a(y+nd)/\pi}(b)]$

$$=(2\pi/ad)\sum_{m_1}^{m_2} H(m)\cos\{b\cos[(\pi/2ad)(2\pi m+x)]\}$$

$\sum_{-\infty}^{\infty} \exp(inx)[\cos(ay+and)]^{-1}[E_{2a(y+nd)/\pi}(b)+E_{-2a(y+nd)/\pi}(b)]$

$$=(-2\pi/ad)\sum_{m_1}^{m_2} H(m)\sin\{b\cos[(\pi/2ad)(2\pi m+x)]\}$$

5.32 $\sum_{-\infty}^{\infty} \exp(inx)(y+nd)^{-1-\nu}\mathbf{H}_\nu[a(y+nd)]$

$$=(2/d)(2\pi/a)^{\frac{1}{2}}\sum_{m_1}^{m_2} H(m)\{a^2-[(2\pi m+x)/d]^2\}^{\frac{1}{2}\nu+\frac{1}{4}}P_{\nu-\frac{1}{2}}^{-\nu-\frac{1}{2}}[|(2\pi m+x)/ad|]$$

$\mathrm{Re}\,\nu>-\frac{3}{2};\quad \mathrm{Re}\,\nu>-\frac{1}{2}$ if $x=\pm ad$

5.33 $\sum_{-\infty}^{\infty} \exp(inx)(y+nd)^{-\mu-1}s_{\mu,\nu}[a(y+nd)]$

$$=(2^{\frac{3}{2}+\mu}\pi^{\frac{1}{2}}/a^{\frac{1}{2}}d)\,\Gamma[\tfrac{1}{2}(\mu+\nu+3)]\,\Gamma[\tfrac{1}{2}(\mu-\nu+3)]\,[(\mu+1)^2-\nu^2]^{-1}$$

$$\cdot\sum_{m_1}^{m_2} H(m)\{a^2-[(2\pi m+x)/d]^2\}^{\frac{1}{2}\mu+\frac{1}{4}}P_{\nu-\frac{1}{2}}^{-\mu-\frac{1}{2}}[|(2\pi m+x)/ad|]$$

$\mathrm{Re}\,\mu>-\frac{3}{2};\quad \mathrm{Re}\,\mu>-\frac{1}{2}$ if $x=\pm ad$

5.34 $\sum_{-\infty}^{\infty} J_{\mu-n}(a)J_{\nu+n}(a)\exp(inx)=\exp[i(\mu-\nu)\frac{1}{2}x]J_{\nu+\mu}(2a\cos\frac{1}{2}x)$

$-\pi\leq x\leq\pi,\quad \mathrm{Re}(\nu+\mu)>0$

5.35 $\sum_{-\infty}^{\infty} J_{\nu-n}(z)J_n(z)\exp(inx)=\exp(i\nu\frac{1}{2}x)J_\nu(2z\cos\frac{1}{2}x)$

5.36 $\sum_{-\infty}^{\infty} \exp(inx)J_n(a)J_{\nu+n}(b)=[(b-ae^{-ix})/(b-ae^{ix})]^{\frac{1}{2}\nu}J_\nu[(a^2+b^2-2ab\cos x)^{\frac{1}{2}}]$

5.37 $\sum_{-\infty}^{\infty} (-1)^n I_n(a)I_{\nu+n}(b)\exp(inx)=[(b-ae^{-ix})/(b-ae^{ix})]^{\frac{1}{2}\nu}I_\nu[(a^2+b^2-2ab\cos x)^{\frac{1}{2}}]$

5.38 $\sum_{-\infty}^{\infty} \exp(inx)Y_{\nu+n}(b)J_n(a)=[(b-ae^{-ix})/(b-ae^{ix})]^{\frac{1}{2}\nu}Y_\nu[(a^2+b^2-2ab\cos x)^{\frac{1}{2}}]$

5.39 $\sum_{-\infty}^{\infty} I_n(a)K_{\nu+n}(b)\exp(inx)=[(b-ae^{-ix})/(b-ae^{ix})]^{\frac{1}{2}\nu}K_\nu[(a^2+b^2-2ab\cos x)^{\frac{1}{2}}]$

5.40 $\sum_{-\infty}^{\infty} a^{n-\mu}J_{\mu-n}(a)b^{-\nu-n}J_{\nu+n}(b)\exp(inx)=\exp[i(\mu-\nu)\frac{1}{2}x][2\cos\frac{1}{2}x/(a^2e^{\frac{1}{2}ix}+b^2e^{-\frac{1}{2}ix})]^{\frac{1}{2}\nu+\frac{1}{2}\mu}$

$\cdot J_{\mu+\nu}\{[2\cos\frac{1}{2}x(a^2e^{\frac{1}{2}ix}+b^2e^{-\frac{1}{2}ix})]^{\frac{1}{2}}\},\quad -\pi\leq x\leq\pi,\quad \mathrm{Re}(\nu+\mu)>0$

5.41 $\sum_{1}^{\infty} [\tau_{\nu,n}J_{\nu+1}(\tau_{\nu,n})]^{-1}J_\nu(\tau_{\nu,n}x)=\frac{1}{2}x^\nu,\quad 0\leq x<1$

5.42 $\sum_{1}^{\infty} [\tau_{0,n}J_1(\tau_{0,n})]^{-2}J_0(X\tau_{0,n})J_0(\tau_{0,n}x)=-\frac{1}{2}\log X,\quad 0\leq x\leq X\leq 1$

5.43 $\sum_{1}^{\infty} \tau_{0,n}^{-\mu-1}[J_1(\tau_{0,n})]^{-2}J_{\mu+1}(\tau_{0,n})J_0(\tau_{0,n}x)=[2^{-\mu-1}/\Gamma(1+\mu)](1-x^2)^\mu$

$0<x<1,\quad \mathrm{Re}\mu>-1$

5.44 $\sum_{1}^{\infty} \tau_{\nu,n}^{-\mu-1}[J_{\nu+1}(\tau_{\nu,n})]^{-2}J_{\nu+\mu+1}(\tau_{\nu,n})J_\nu(\tau_{\nu,n}x)=[2^{-\mu\ 1}/\Gamma(1+\mu)]x^\nu(1-x^2)^\mu$

$0<x<1,\quad \mathrm{Re}\nu>-1$

5.45 $\sum_{1}^{\infty} \tau_{\nu,n}(\tau_{\nu,n}^2-z^2)^{-1}[J_\nu(\tau_{\nu,n}x)/J_{\nu+1}(\tau_{\nu,n})]=\frac{1}{2}[J_\nu(xz)/J_\nu(z)], \quad 0\leq x<1$

5.46 $\sum_{1}^{\infty} (\tau_{\nu,n}^2-z^2)^{-1}\tau_{\nu,n}^{-1}[J_\nu(\tau_{\nu,n}x)/J_{\nu+1}(\tau_{\nu,n})]=\frac{1}{2}z^{-2}\{[J_\nu(xz)/J_\nu(z)]-x^\nu\}, \quad 0\leq x<1$

5.47 $\sum_{1}^{\infty} \tau_{\nu,n}(\tau_{\nu,n}^2+z^2)^{-1}[J_\nu(\tau_{\nu,n}x)/J_{\nu+1}(\tau_{\nu,n})]=\frac{1}{2}[I_\nu(xz)/I_\nu(z)], \quad 0\leq x<1$

5.48 $\sum_{1}^{\infty} (z^2-\tau_{\nu,n}^2)^{-1}[J_{\nu+1}(\tau_{\nu,n})]^{-2}J_\nu(\tau_{\nu,n}X)J_\nu(\tau_{\nu,n}x)$

$=[\pi J_\nu(xz)/4J_\nu(z)][J_\nu(x)Y_\nu(Xz)-J_\nu(Xz)Y_\nu(z)]$

$=[\pi i J_\nu(xz)/4J_\nu(z)][J_\nu(z)H_\nu^{(2)}(Xz)-J_\nu(Xz)H_\nu^{(2)}(z)]$

5.49 $\sum_{1}^{\infty} (z^2+\tau_{\nu,n}^2)[J_{\nu+1}^2(\tau_{\nu,n})]^{-2}J_\nu(\tau_{\nu,n}X)J_\nu(\tau_{\nu,n}x)$

$=[I_\nu(xz)/2I_\nu(z)][I_\nu(z)K_\nu(Xz)-K_\nu(z)I_\nu(Xz)]$

In the last two formulas $0\leq x\leq X\leq 1$. If $0\leq X\leq x\leq 1$, interchange x and X at the right sides.